The Microbiology of Water 1994
Part 1—Drinking Water

Report on Public Health and Medical Subjects No. 71

Methods for the Examination of Waters and Associated Materials

London: HMSO

First published 1994

ISBN 0 11 753010 7

HMSO

Standing order service

Placing a standing order with HMSO BOOKS enables a customer to receive other titles in this series automatically as published.

This saves the time, trouble and expense of placing individual orders and avoids the problem of knowing when to do so.

For details please write to HMSO BOOKS (PC 11B/2), Publications Centre, PO Box 276, London SW8 5DT quoting reference X22.04.22.

The standing order service also enables customers to receive automatically as published all material of their choice which additionally saves extensive catalogue research. The scope and selectivity of the service has been extended by new techniques, and there are more than 3,500 classifications to choose from. A special leaflet describing the service in detail may be obtained on request.

Foreword

This volume (the sixth edition) is the revision of the 1983 DoE/DH/PHLS publication 'The Bacteriological Examination of Drinking Water Supplies 1982'.

Whilst the main aim of this edition is to provide users with up to date methods for the microbiological examination of drinking water, additional information on the regulatory background, sampling, quality assurance and statistical considerations, and health and safety requirements are included.

This revision also addresses a change in the definition of coliform organisms. In the past, the definition was based on the behaviour of organisms in lactose fermentation methods traditionally used for coliform isolation. This presented a barrier to the development of new and improved methodologies since it ultimately required the demonstration of the production of gas from lactose during fermentation. When 'Guidance on Safeguarding the Quality of Public Water Supplies' was published in 1989 it contained a definition of coliform organisms based on the possession of genes necessary to break down lactose, namely the gene for ß-galactosidase. This revision confirms that definition and removes the requirement for the demonstration of gas production as a necessary characteristic of coliform organisms. The use of a genetically based definition opens up possibilities for the use of new techniques, provided the performance is demonstrated to be equal to, or better than, the performance of methods described in this volume.

Members of the committee assisting with this booklet

Grateful acknowledgement is made to all members and colleagues who contributed draft manuscripts and to those who made comments and helpful suggestions.

K Aston		Thames Water Utilities Ltd
R A E Barrell	1,2	PHLS (Manchester)
C Benton	1,2	Strathclyde Water
C R Chada	1,2 (from June 1990)	DOE (Northern Ireland)
J S Colbourne	1,2	Thames Water Utilities Ltd
S Cole	2	Wessex Water
J V Dadswell	1,2	PHLS (Reading)
A Dawson	1,2 (to October 1993)	Department of Health
P A Deacon	1,2	Consultant
P J Dennis		Thames Water Utilities Ltd
C Hill	1 (to December 1992)	Drinking Water Inspectorate
J A M Hilton	2 (from October 1993)	Department of Health
J V Lee	1,2 (from June 1991)	PHLS (Nottingham)
M J Lewis	1,2 (to May 1993)	PHLS (Nottingham)
N F Lightfoot	2 (from November 1989)	PHLS (Newcastle)
S L Mawer		PHLS (Hull)
E B Pike	1 (from March 1991)	WRc plc
L R Pittwell	1,2 (to April 1991)	Drinking Water Inspectorate
B Pound	2	Severn Trent Water Ltd
G Stanfield	1 (to March 1991)	WRc plc
H E Tillett	1,2	PHLS (Colindale)
K C Thompson		Yorkshire LabServices
W M Waite		Drinking Water Inspectorate
J Watkins	1,2 (from March 1991)	Yorkshire LabServices
P Wyn-Jones		Sunderland University

1 Member of panel 2.5 (Report 71) responsible for drafting this report
2 Member of working group 2

Grateful acknowledgement is also made to the following organisations who commented on the final draft.

Association of Public Analysts
Department of Health
Public Health Laboratory Service
Scottish Association of Directors of Water and Sewerage Services
Water Companies' Association
WRc plc
Water Services Association

Contents

About this series

Introduction

This booklet is part of a series intended to provide authoritative guidance on recommended methods of sampling and analysis for determining the quality of drinking water, groundwater, river and seawater, waste water and effluents as well as sewage sludges, sediments and biota. In addition, short reviews of the more important analytical techniques of interest to the water and sewage industries are included.

Performance of methods

Ideally, all methods especially for chemical parameters should be fully evaluated with results from performance tests reported. These methods should be capable of establishing, within specified or pre-determined and acceptable limits of deviation and detection, whether or not any sample contains concentrations of parameters above those of interest.

For a method to be considered fully evaluated, individual results encompassing at least ten degrees of freedom from at least three laboratories should be reported. The specifications of performance generally relate to maximum tolerable values for total error (random and systematic errors), systematic error (bias), total standard deviation and limit of detection. It is recognised that the performance criteria expected for chemical parameters will not be strictly applicable to microbiological methods. Often, full evaluation is not possible and only limited performance data may be available.

In addition, good laboratory practice and analytical quality control are essential if satisfactory results are to be achieved.

Standing Committee of Analysts

The preparation of booklets in the series 'Methods for the Examination of Waters and Associated Materials' and their continuous revision is the responsibility of the Standing Committee of Analysts. This committee was established in 1972 by the Department of the Environment and is managed by the Drinking Water Inspectorate. At present there are nine working groups, each responsible for one section or aspect of water quality analysis. They are:

1.0 General principles of sampling and accuracy of results
2.0 Microbiological methods
3.0 Empirical and physical methods
4.0 Metals and metalloids
5.0 General non-metallic substances
6.0 Organic impurities
7.0 Biological monitoring
8.0 Sewage works control methods
9.0 Radiochemical methods

The actual methods and reviews are produced by smaller panels of experts in the appropriate field, in co-operation with the working group and main committee. The names of those members and organisations associated with this booklet are listed at the front.

Publication of new or revised methods will be notified to the technical press. An index of methods and the more important parameters and topics is available from HMSO (ISBN 0 11 752669 X).

Every effort is made to avoid errors appearing in the published text. If however, any are found, please notify the Secretary.

Dr D WESTWOOD
Secretary

14 May 1994

Warning to users

The analytical procedures described in this booklet should only be carried out under the proper supervision of competent, trained analysts in properly equipped laboratories.

All possible safety precautions should be followed and appropriate regulatory requirements complied with. This should include compliance with The Health and Safety at Work etc Act 1974 and any regulations made under the Act, and the Control of Substances Hazardous to Health Regulations 1988 SI 1988/1657. Where particular or exceptional hazards exist in carrying out the procedures described in this booklet then specific attention is noted. Numerous publications are available giving practical details on first aid and laboratory safety and these should be consulted and be readily accessible to all analysts. Amongst such publications are those produced by the Royal Society of Chemistry, namely 'Safe Practices in Chemical Laboratories' and 'Hazards in the Chemical Laboratory, 5th edition, 1992'; by Member Societies of the Microbiological Consultative Committee, 'Guidelines for Microbiological Safety, 1986, Portland Press, Colchester'; and by the Public Health Laboratory Service 'Safety Precautions, Notes for Guidance'. Another useful publication is produced by the Department of Health entitled 'Good Laboratory Practice'.

Preface

The importance for public health of the provision of a wholesome supply of drinking water has been recognised since the 19th century. Knowledge, understanding and good practice have developed over time with consequent benefits to public health. Today the maintenance of a sufficient supply of wholesome drinking water is a complex undertaking in which individuals from many disciplines take part. This report aims to assist those individuals in their work to maintain the supply of microbiologically wholesome drinking water.

The first Report on "The Bacteriological Examination of Water Supplies", was the first in a series of Ministry of Health publications on Public Health and Medical Subjects. This Report was prepared in 1934 by a small Ministry of Health committee under the Chairmanship of the late Dr Thomas Carnworth with the help of the late Sir Alexander Houston and representatives of the Lister Institute of Preventive Medicine, the London School of Hygiene and Tropical Medicine and the Counties' Public Health Laboratories. In 1956 the Public Health Laboratory Service (PHLS) Water Committee assumed responsibility for revising the Report. In 1973, the Department of the Environment (DoE) became responsible for all aspects of the water cycle and established the Standing Committee of Analysts (SCA) to review and keep up to date the methods recommended for water examination in the United Kingdom. The fifth edition, issued in 1983, was therefore produced under the auspices of the Standing Committee of Analysts and published jointly by DoE, Department of Health and Social Security (DHSS) and PHLS.

A further revision of the Report has become necessary as a result of major changes in the United Kingdom water industry, including the passing of the Water Act 1989 now consolidated into the Water Industry Act 1991 (the Act), the Water Supply (Water Quality) Regulations 1989 (the Regulations), and the Private Water Supplies Regulations 1991 (the Private Supplies Regulations). This revision has been produced by Working Group 2 of the DoE Standing Committee of Analysts.

The new edition is intended to be the first in a series of publications dealing with all aspects of the microbiology of water. As in previous editions, the main section is devoted to reference methods for the isolation, enumeration and identification of micro-organisms found in water supplies that are considered to be of significance, either for sanitary or other reasons. For the first time, tentative methods are included. These are considered to be capable of giving useful results but require further evaluation before achieving fully validated status. Other new sections deal with microbiological aspects of source water quality, water treatment and distribution, and statistical considerations, quality assurance and an expanded section on laboratory safety.

Practical advice on water quality matters is included in "Guidance on Safeguarding the Quality of Public Water Supplies" (the Guidance Document) and its Scottish equivalent. Definitions of relevant organisms are recommended which depart from those traditionally used by water microbiologists. With respect to coliform organisms, the demonstration of the production of gas is now no longer considered relevant. Definitions are more scientifically based and will include organisms of possible sanitary significance that would previously have been excluded by the former 'method-related' definitions. Being no longer 'method-related' they will enable immunological, genetic and electrometric techniques to be used for the detection and confirmation of coliforms provided these methods are shown to be of equivalent or better performance than those described here.

The maintenance of microbiologically wholesome drinking water supplies requires the commitment of individuals from many different disciplines: professionals in water undertakers, consultants in communicable disease control, environmental health officers, public analysts and hospital and PHLS microbiologists. Good communication and liaison between all these individuals is essential to enable appropriate action to be taken whenever water quality problems occur.

In considering the public health aspects of water supplies, local authorities should consult their medical advisors. In most cases this will be the Consultant in Communicable Disease Control (CCDC) (in Scotland, the Consultant in Public Health Medicine) who also acts as Proper Officer for the Local Authority. Where this is not the case, the local authority should consult the Director of Public Health (DPH) for the district health authority (DHA). Revised guidance on the role of the CCDC was issued by the Department of Health (DH) in November 1993 (HSG(93)56). Advice was issued by the Welsh Office to the Welsh authorities in April 1989 (WHC(89)1). It is emphasised that it is the responsibility of the CCDC (or, in the absence of a CCDC, the DPH) to give medical advice if there is any microbiological evidence to suggest that a water supply may be unwholesome. If there is evidence that the water is unsafe to drink, it is the duty of the CCDC to advise immediately on any further action and investigations considered necessary. If water-borne infections are suspected, the CCDC should assume executive responsibility and a co-ordinating role for the control of any outbreak of disease. The CCDC should usually seek advice from the local public health laboratory or its equivalent and, if necessary, the assistance of the regional epidemiologist and the PHLS Communicable Disease Surveillance Centre (CDSC). HSG(93)56 makes it clear that suitable local arrangements at all levels should be made to ensure that there is direct and continuing liaison between all those concerned with communicable disease control; this includes water undertakers, local authorities, DHA and other National Health Service (NHS)authorities and the PHLS. The importance of clear lines of communication, both within and between these organisations, cannot be over-emphasised.

The Regulations apply to all water from public supplies for drinking, cooking, washing or food production. The same standards for wholesomeness are included in the Private Supplies Regulations. Although water suppliers are required to ensure the wholesomeness of their product, it is well recognised that it is difficult to exercise control over changes in water quality to all premises in their areas. Local authorities are required to take samples from private water supplies and have the power to take samples from public water supplies.

As well as giving details of approved methods, this Report gives general advice on many microbiological aspects connected with potable water supplies. The microbiological safety of these supplies in the United Kingdom has been assured in no small measure by regular monitoring and observance of the recommendations contained in previous editions of Report 71. It is hoped that this revised issue will be found to be even more useful than before, not only in the United Kingdom but also internationally.

Chapter 1: Introduction

This Report, recognising the importance to public health of the proper maintenance of a wholesome supply of drinking water, has been prepared with the following objectives:

—to outline the principles on which the microbiological examination of drinking water is based;

—to recommend techniques for sampling and examination to ensure good scientific practice and comparability of results; and

—to give advice on the interpretation of the microbiological results and on the need for consequent action in the event of failure of microbiological standards.

Almost the whole of the population of the United Kingdom is served by a piped public supply of treated drinking water and less than 1% of the population are served by private supplies. This Report is primarily concerned with the microbiological examination and monitoring of public water supplies—not only as the water is distributed, but also as it is abstracted and treated. Mention is also made of risks which might arise from inadequate treatment, especially following a sudden pollution of the source, as well as to contamination within the distribution system. In addition, since the Regulations apply to all water for drinking purposes, brief reference is made in this Report to the microbiological examination of particular supplies, such as those on trains, ships and aircraft, and in hospitals, food establishments and similar premises.

A single laboratory examination of any water does not justify the conclusion that all is well and that the supply will remain suitable for drinking purposes. Contamination is often intermittent and may not be revealed by the examination of a single sample. The impression of security given by microbiological testing of a water at infrequent intervals may, therefore be quite false. Indeed the value of microbiological tests is dependent upon their frequent and regular use. **It is far more important to examine a supply frequently by a simple test than occasionally by a more complicated test or series of tests**. Information gained in the course of time through monitoring will provide a comprehensive picture of the range of quality at any particular source of water, any lapse from which must at once arouse suspicion. The most a microbiological report based on a single sample can prove is that at the time of examination, certain bacteria indicative of faecal contamination did or did not grow under laboratory conditions from the sample of water as received and tested. It must be emphasised that, when site inspection shows a water supply to be obviously subject to contamination, remedial action should be taken without waiting for, and irrespective of, the results of microbiological examination.

The European Community Directive relating to the Quality of Water Intended for Human Consumption, Directive 80/778/EEC (the Directive), lays down standards for the quality of drinking water and water for use in food production undertakings. The Water Act 1989 (now consolidated in the Water Industry Act 1991, which came into force on 1st December 1991), the Water Supply (Water Quality) Regulations 1989 and the Private Water Supplies Regulations 1991 transpose the Directive into the law of England and Wales. Similar legislation has been introduced in Scotland and is in the course of introduction in Northern Ireland.

The Act places a duty on water undertakers to supply only water that is wholesome at the time of supply. The time of supply is the moment when water passes from the company's pipe into pipes owned by the owner of premises or property. Companies are not responsible for a deterioration of drinking water quality which occurs within consumers' premises, although water undertakers are required to take certain steps if there is a risk that the standards for lead, copper or zinc might be breached as a result of corrosion of the consumers' pipes or fittings. Wholesomeness is defined by reference to the standards and other requirements of the Regulations. The Act also creates a criminal offence of supplying water that is unfit for human consumption. Under the Act, the Secretary of State for the Environment and the Secretary of State for Wales are empowered to appoint technical assessors. The Drinking Water Inspectorate came into being on 2nd January 1990 to fulfil that function. The Inspectorate advises the Secretaries of State as to whether water companies are supplying wholesome water and are monitoring, recording and reporting on the quality of drinking water in accordance with legal requirements.

The Regulations originally applied only to water supplied for domestic purposes (that is, drinking, washing and cooking). Following amendments made to the Act by the Food Safety Act 1990 (the Food Act), the Regulations have been amended, and now also apply to water supplied for food production purposes. Prescribed concentrations or values (PCV) are specified for microbiological, chemical and physical parameters. Mandatory sampling frequencies for treatment works and water supply zones are specified according to the volume of water supplied or the population served. In addition, weekly monitoring for microbiological parameters is specified for service reservoirs and water towers in use. Regulations under the Food Act prescribe standards for drinking water supplied in sealed containers, which includes bottles (see Appendix E).

Private water supplies are defined in the Act as any supplies of water provided otherwise than by a statutorily appointed water undertaker. The definition includes water abstracted from a source on the premises on which it is used or consumed, bulk supplies purchased from a water undertaker and subsequently redistributed, and also sources or supplies used to produce bottled water. The Private Supplies Regulations set standards of wholesomeness and specify the duties of local authorities in respect of sampling and analysis of private water supplies. Where a supply is unwholesome or insufficient the Act allows local authorities to serve a notice requiring either improvements to the supply or connection of the premises to a public supply. The Department of the Environment has issued Circular 24/91 and the Welsh Office, Circular 68/91, which provides guidance to local authorities on their duties to sample and analyse private water supplies. Similar guidance has been issued by the Scottish Office Environment Department (Circular 20/1992). Advice on treatment has been given in 'Manual on Treatment of Private Water Supplies' (DoE/WO/SO 1993). The quality of natural mineral water is covered by the Natural Mineral Waters Regulations 1985 (SI 1985/71 as amended by SI 1990/2487) (see Appendix E).

Chapter 2: Public Health Aspects

2.1 Water-borne Infections

The microbiological safety of potable water depends upon the exclusion of pathogenic micro-organisms that cause infections via the gastro-intestinal tract. The recognition in the 19th century that cholera and typhoid fever could be spread by the ingestion of water contaminated with excrement laid the foundations of present day treatment processes and piped distribution, the design and operation of which must always ensure that potable water cannot transmit such infections. Other treatment objectives should never take precedence over this primary requirement.

2.1.1 Water-borne pathogens

Many micro-organisms have now been recognised as capable of being transmitted by water, including bacteria, protozoa and viruses. Their essential properties are that they are present in large numbers in human or animal excreta or both, and are relatively resistant to environmental decay. Many are highly likely to cause infections if ingested in small numbers. Examples of those known to have caused water-borne infections are listed in Table 1.

Table 1 Known Water-borne Pathogens

Bacteria

Campylobacter species
Escherichia coli (certain serotypes)
Salmonella species (including *S.typhi*)
Shigella species
Streptobacillus moniliformis
Vibrio species (including *V.cholerae*)
Yersinia enterocolitica

Protozoa

Balantidium coli
Cryptosporidium species
Entamoeba histolytica
Giardia intestinalis (= *lamblia*)

Viruses

Hepatitis A virus
Hepatitis E virus
Small round structured viruses
(for example Norwalk virus)
Rotaviruses

Other organisms causing enteritis are known to be present in human and animal excreta and therefore have the potential to cause water-borne infections. These include the aeromonas group, *Bacillus cereus*, certain clostridia, *Plesiomonas shigelloides*, *Isospora* species and some enteric viruses. Certain enteroviruses belonging to the poliovirus and coxsackievirus groups have occasionally been isolated in small numbers from drinking water but there is no evidence that illness has resulted; direct person-to-person transmission being the usual route. A more detailed consideration of viruses in the water cycle is given in 'Methods for the isolation and identification of human enteric viruses from water and associated materials' (SCA 1994) (see section 3.7).

Organisms normally present in the environment, such as saprophytic mycobacteria, *Pseudomonas aeruginosa* and some aeromonads, have the potential to cause infections in immuno-compromised people and have at times been suspected of being transmitted by the water route. *Listeria monocytogenes* is widely distributed in the environment and there is a possibility that it might be found in water. However, while probable food-borne listeriosis has been noted there is no suggestion of transmission by drinking water. Infections caused by *Legionella pneumophila*, *Naegleria fowleri* and *Acanthamoeba* are associated with warm water but require special circumstances to enable infection to occur, notably direct inhalation of water or aerosols. Thus infection is not directly attributable to drinking water supplies.

Most enteric pathogens are found in the intestinal tract of persons suffering from acute infections caused by organisms which are present in large numbers in excreta. Some infections, such as typhoid fever, may produce a prolonged carrier state after the acute phase is over, and the organisms can be excreted in relatively small numbers for many years. Organisms derived from these sources will then be present in sewage.

Animals may also harbour in their gut organisms pathogenic for humans and may therefore be a significant source of pollution. Gulls pose a particular problem on catchments, at refuse tips and sewage treatment works. Modern intensive farming methods result in the production of large quantities of slurry which may subsequently contaminate watercourses from which water is abstracted for drinking. This is thought to have been the cause of an outbreak of water-borne cryptosporidiosis reported in 1989 (Richardson et al 1991).

In addition to the drinking of contaminated water, its use in the preparation of food, which may allow the multiplication of enteric pathogens, presents obvious dangers.

The routes by which enteric pathogens can gain access to drinking water are numerous. Contamination of catchment areas will result in the presence of pathogens in surface waters from which drinking water may be abstracted. Less obviously, ground water may also be contaminated.

Outbreaks of water-borne infection have usually resulted from defects in one or more treatment processes (or lack of any treatment) or because there has been ingress of pathogens as a result of activities causing damage to some part of the distribution system. Defects in plumbing, cross-connections, back flow and pressure fluctuations can cause contamination of distribution systems, and can result in the transmission of organisms causing infection.

2.1.2 Outbreaks of water-borne infections

The recognition of an outbreak of water-borne infection will, in part, depend upon the continuing surveillance of the prevalence of communicable infections in a particular locality. This will depend initially on the observations by general practitioners or by the local CCDC of a sudden increase in the incidence of infections that could be water-borne. The monitoring by water undertakers of consumer complaints is also important and helpful. Sometimes the CCDC or Environmental Health Officer (EHO) may receive hearsay information suggesting the possibility of a water-borne infection. Other early intimations may arise from an increase in possible water-borne infections (noted by a diagnostic microbiology laboratory) being reported to the CCDC and/or the CDSC, or in Scotland the Communicable Disease and Environmental Health (Scotland) Unit (CD(S)U). Any suspicion of a water-borne outbreak should be notified immediately to the water undertaker, with whom close liaison is essential. If a water-borne outbreak is confirmed, the water undertaker must notify the Drinking Water Inspectorate, whilst the health authority should notify the Department of Health. If a private supply is implicated the primary responsibility for investigation and control lies with the local authority and the DHA in the affected area. The local water undertaker, while not directly involved, should be made aware and can often provide technical help and advice.

The possibility that an outbreak of water-borne infection may result from operational problems such as a failure of disinfection, filter breakdown or a major pollution incident must always be considered by water undertakers, and in these circumstances the local and district health authorities in the area of the supply must be notified as soon as possible.

The National Rivers Authority (NRA) in England and Wales, and in Scotland the River Purification Authorities, have a statutory responsibility for maintaining and upgrading the quality of surface waters and for the protection of groundwater supplies. As all public water supplies are derived from these sources, water undertakers must be informed of any deterioration in quality and pollution incidents which may result in the need for temporary cessation of abstraction or urgent modification of the treatment processes (or other action) at a treatment works.

Close and continuing liaison between water undertakers, district health authorities and local authorities is essential to ensure that appropriate action can be taken to investigate and control outbreaks of water-borne infectious diseases in the local community. The Group of Experts on Cryptosporidium in Water Supplies, (Badenoch Group), (DoE/DH 1990) recommended that district health authorities, local authorities and water companies should together draw up an Outbreak Control Plan (OCP) for the control of water-borne infection in the community, and that this plan should provide for the establishment of an Outbreak Control Team (OCT) to be called into being by the DPH or the CCDC should an outbreak be suspected. The OCT must include representatives from the district health authority, local authority and water undertaker.

Depending on the size of the outbreak, assistance may need to be sought from the regional epidemiologist and from CDSC, who should in any case be kept informed. The assistance of the nearest public health laboratory may be required if there are limited facilities in local NHS or other laboratories.

In the event of an outbreak of water-borne disease, the OCT will be responsible for ensuring that appropriate action is taken without delay. Arrangements will depend on the nature and size of the incident and may require the provision of an alternative supply of drinking water or the issuing of advice to boil the water until the source of the infection has been identified and eliminated. When the need to issue advice to boil water is being assessed the requirements of others, including food, dairy and pharmaceutical industries, dentists and veterinarians and also patients in hospitals or on home dialysis, will need to be considered. It is essential that there is close liaison between district health and local authorities and water undertakers to ensure that clear meaningful advice is given to all persons and organisations likely to be affected. The media can assist with communication of this advice.

The importance of a clear system of emergency communication within the water undertaking between the scientific, managerial and operational staff is self-evident. This will ensure that any communication with consumers and the media, and any remedial action that may be required at the treatment works or in the distribution system can be instituted without delay. General guidance on the measures that may be required is given in 'Guide to the microbiological implications of emergencies in the water services' (WAA 1985) and 'Operational guidelines for the protection of drinking water supplies' (WAA 1988), both originally published by the Water Authorities Association (now the Water Services Association), while the wider public health implications are addressed in detail in 'Drinking Water Supplies: a Microbiological Perspective' (DH 1993a).

Chapter 3: Microbiological Examination of Drinking Water: Rationale

Full examination of a water supply embodies several lines of investigation including topographical, chemical, biological and microbiological (of which bacteriology is crucially important) and organoleptic characterization. Each line of investigation has its uses and indications. This report deals mainly with bacteriogical aspects, but other microbiological aspects are also covered.

Bacteriological examination is particularly important because it remains the most sensitive method of detecting faecal and therefore potentially dangerous pollution.

For several reasons, monitoring for the presence of specific pathogenic bacteria, viruses and other agents in water is impracticable and indeed unnecessary for routine control purposes. Any pathogenic micro-organisms present in polluted water are usually greatly outnumbered by the normal commensal bacterial flora of the human or animal intestine. Although it may be possible to isolate microbial pathogens from contaminated water, especially when it is heavily polluted, large volumes (several litres) of the water may need to be examined. Selective media and special equipment may also be required for the isolation, and subsequent identification of organisms may involve biochemical, serological and/or other tests. Reliance is therefore placed on relatively simple and more rapid tests for the detection of certain commensal intestinal bacteria, especially *Escherichia coli* and other coliform organisms, because they are easier to isolate and characterise and because they are almost always present in the faeces of humans and warm-blooded animals, and hence in sewage, in large numbers. The presence of such faecal indicator organisms in a sample of drinking water thus denotes that intestinal pathogens could be present, and that the supply is therefore potentially dangerous to health (Waite 1991). The absence of these faecal indicator organisms, however, cannot be taken as an absolute indication that intestinal pathogens are also absent.

Chemical analysis, though lacking the sensitivity of bacteriology in this respect, is nevertheless an important aid to the hygienic assessment of a water supply; however, its major roles are to provide process control information for water treatment and to monitor for compliance with the prescribed standards. Biological examination is used to detect the presence of algae and animal life which may impede performance of filtration and gain access to supplies through suboptimal water treatment or defects in the distribution network. Topographical examination of catchment areas and water supply networks may reveal potential risks which would pass unnoticed by any other method.

3.1 Organisms Indicative of Faecal Pollution

The search for organisms indicative of faecal pollution instead of for pathogens themselves is universally accepted for monitoring the microbial pollution of water supplies. Ideally, the detection of these indicator organisms should denote the potential presence of intestinal pathogens. Indicator organisms should be abundant in faeces and sewage, and be absent or at least occur in very small numbers in all other sources. They should be capable of easy isolation, identification and numerical estimation and unable to grow in the aquatic environment. In practice there is no organism which consistently meets all these criteria, but in temperate climates most of them are fulfilled by the bacterium *Escherichia coli*. This organism is reaffirmed as the essential indicator of pollution by faecal material of human or animal origin. Other organisms which possess some of these properties can also be used to provide supplementary information in certain circumstances. They include other coliform organisms, faecal streptococci, *Clostridium perfringens* and possibly other intestinal commensals. The particular test or combination of tests to be used in water examination must be left to the discretion of the microbiologist, as they will depend on the nature of the sample, the actual circumstances and the information required.

The Regulations specify numerical standards for total and faecal coliforms, faecal streptococci and sulphite-reducing clostridia, and descriptive standards for total viable counts at 37°C and 22°C. Routine monitoring is not specified for faecal streptococci or sulphite-reducing clostridia. Examination for other organisms should be undertaken when it is considered on epidemiological or other grounds that they might be present and thereby be rendering the water unwholesome.

3.2 Definitions

Early water microbiologists recognised that faeces contain large numbers of bile salt-tolerant, facultatively anaerobic organisms capable of growing at 37°C and of fermenting lactose at this temperature, and so began to enumerate these organisms as 'indicators of faecal pollution'. The method generally employed was inoculation of MacConkey broth with incubation at 37°C for 48 hours, followed by subculture from cultures producing acid and gas to MacConkey agar to eliminate obligate anaerobes. The organisms isolated in this manner were referred to as "coliform organisms". As a consequence, the definition of coliform organism adopted was essentially that of *any* organism isolated by this method. Subsequent recognition of the existence of coliform organisms which were not always of faecal origin led to the introduction of differential tests to permit the separate characterisation of coliforms able to grow and ferment lactose at 44°C with the production of gas and to produce indole from tryptophan. These were considered to be exclusively of faecal origin and were called *Bacterium coli* Type 1, later to be renamed *Escherichia coli* (*E. coli*).

Definition of coliforms in terms of the method generally used for their detection was perfectly acceptable as long as methodology remained unchanged. When alternative methods began to emerge which do not rely on the production of acid and gas from lactose, the continued use of method-related definitions necessitated the adoption of confirmatory tests to verify gas production from lactose in liquid media.

Production of detectable volumes of gas from lactose within 48 hours at 37°C is not a characteristic which can in general be used to differentiate between species of the Family Enterobacteriaceae. Within most lactose-fermenting species strains occur, and may be common, which do not produce gas from lactose under these conditions. Indeed, some 5% of clinical isolates of *E. coli* from faeces have, in some studies, been anaerogenic at 37°C. There is no suggestion that the production of gas is in any way related to the sanitary significance of lactose-fermenting enterobacteria. However, if a coliform is to have been of faecal origin it must have been capable of growth at 37°C.

Within the Family Enterobacteriaceae, the fermentation of lactose invariably begins with the cleavage of lactose into glucose and galactose by the enzyme ß-galactosidase. Since the concept of the coliform is of strains of Enterobacteriaceae able to ferment lactose it follows that the most fundamental characteristic of a coliform must be the possession of a gene coding for the production of ß-galactosidase. The expression of this gene can be affected by many factors such as time, temperature and medium. Under different conditions the same organism may or may not ferment lactose sufficiently to register as a lactose-fermenter under the test conditions in use. The production of gas from lactose is a particularly variable characteristic. For these reasons new definitions of coliforms and *E. coli* are proposed (see section 7.7). These definitions should be regarded as practical working definitions for water examination purposes and not taxonomic definitions.

The new definition of coliform bacteria is based on the possession of the ß-galactosidase gene, and seeks to give a standard definition which is not method-related. It should facilitate the development of new operationally useful methods. For the methods described in this Report the presence of the ß-galactosidase gene is demonstrated by the production of acid from lactose. Alternative methods should give performance comparable to or better than that of the reference methods in this Report, and the production of acid from lactose should be the criterion against which alternative methods are to be assessed. It should be appreciated that different methods will demonstrate different populations of coliforms however they are validated. It should be noted that this definition no longer contains the requirement for the demonstration for the production of gas.

3.3 Microbiological Monitoring of Water Supplies

The water microbiologist is faced with the examination of water samples taken from different stages in the treatment process, from the initial water source to the point of use. Current practice assumes in large part that all requirements are equally well met by enumeration of total coliforms and *E. coli*, as simple indicators of faecal pollution. However, the waters to be examined differ widely, and include raw surface and ground waters for abstraction, treated water entering supply, water in distribution and water within premises. In general, two types of indicator are required. These are a faecal indicator which demonstrates the presence and extent of faecal contamination and a quality indicator which can be used to measure the effectiveness of treatment and the integrity of distribution.

Whatever the organism used, monitoring relies on careful scrutiny of counts from samples collected over a period of time. There is no global equation for defining a 'significant' single high count or a change in average count. Good knowledge of the water source, data summaries which include graphs, and interpretation with common sense are important.

3.3.1 Raw waters

3.3.1.1 Surface waters and relatively polluted ground water

The water microbiologist has a clear requirement to use indicators, both of faecal pollution and water quality, when monitoring surface waters to be abstracted for potable supply. Of the traditionally used indicator organisms, *E. coli* is generally regarded as being a suitably specific faecal indicator and should continue to be used as such in this context.

When considering the bacterial load in relation to treatment, the ideal measure would be the total vegetative bacterial cell count, but since treatment does not reliably remove all bacterial spores the total viable count is of limited value. Some microbiologists (Bonde 1977, Mossel 1982) have advocated the use of total Enterobacteriaceae as indicators of process effectiveness and overall quality. However, to maintain historical continuity of data and comply with current legislation and practice, the use of the total coliform count is still recommended as a quality indicator.

3.3.1.2 Relatively unpolluted groundwaters

Many groundwaters are of historically good quality and show little or no evidence of a bacterial population. In such cases the total viable count at 37°C can be the most useful quality indicator, and can provide an early sign of more serious pollution. The total viable count at 22°C has limited public health significance but may perhaps provide useful information as a quality indicator. Additional valuable information can be provided by the examination for total coliforms and *E. coli*.

When total coliforms are detected but are found not to be *E. coli*, examination for faecal streptococci, which survive in water rather longer than coliforms, may help elucidate whether the coliforms were faecal in origin.

Clostridium perfringens spores are capable of surviving for considerable periods in water, and their presence in a groundwater in the absence of other faecal indicators may reveal a connection with surface water or an episode of pollution which had happened some time past. Examination for this organism is therefore worthwhile as part of the examination of a groundwater which has not been regularly monitored in the recent past.

3.3.2 Treated water entering supply

The water microbiologist examining treated water entering supply wishes to monitor treatment effectiveness using a quality indicator which is at least as resistant to the applied treatment as the most resistant pathogen of concern. Since it is known that certain enteroviruses and the cysts of *Giardia intestinalis* are more resistant to chlorine than *E. coli* and that oocysts of *Cryptosporidium* are almost completely resistant, the best quality indicator should be the one which is chlorine-resistant, but capable of being removed to a high degree by treatment before chlorination. Such properties appear to be shown by spores of *Clostridium perfringens* (Payment 1991a,b). Examination for these may be helpful when contamination of treated water with protozoal cysts or oocysts is suspected. This test is not normally in routine use by water microbiologists, and the monitoring for total coliforms is the most valuable test in current use in this context. In general, whether a coliform surviving treatment is of faecal origin or not makes no difference to its significance as an indication that treatment has been inadequate. However, it must be stressed that absence of coliforms cannot guarantee that viruses or protozoal pathogens are also absent.

Any finding of presumptive faecal coliforms must be viewed with concern and demands immediate attention. The implication is that there has been a breach of treatment integrity, with the risk that pathogens may be present.

3.3.3 Water in distribution

Information of interest includes;

(i) faecal ingress in distribution;

(ii) non-faecal ingress in distribution; and

(iii) bacterial growth within distribution.

Examination for a faecal indicator is essential if potentially or actually dangerous ingress has taken or is taking place. The faecal indicator of choice is *E. coli*.

Non-faecal ingress and growth in distribution are not always easily differentiated in practice, but similar information is required in both situations. Enumeration of total coliforms will provide useful information and is indeed required to satisfy the requirements of the Regulations.

The estimation of total viable counts at 37°C and 22°C may generate useful information. To be of value, samples need to be taken regularly from the same site. Random samples required by the Regulations for compliance monitoring provide total viable count information which is more difficult to correlate and interpret.

3.3.4 Water within premises

Once water enters consumers' premises changes in microbiological quality cease to be the responsibility of the water undertaker. Deterioration within premises is discussed in section 4.6.

3.3.5 Rationale for the use of indicator organisms

There is no absolute correlation between the number of indicator organisms and the actual presence or numbers of enteric pathogens, nor between the risk of illness occurring and the numbers of organisms present in a given sample. The finding of *E. coli* in a properly treated water indicates the presence of material of faecal origin and thus a potentially dangerous situation, the nature and extent of which is best determined by immediate 'on-site' and laboratory investigations.

The choice of tests in the detection and confirmation of coliform organisms, including *E. coli*, should be regarded as part of a continuous sequence, the extent for any particular sample depending partly on the nature of the water and partly on the reasons for the examination. Irrespective of their actual identity, all members of the coliform group of organisms as defined may be faecal in origin and an explanation of their presence must always be sought. If further investigation raises doubt as to the faecal nature of the pollution, examination for secondary indicator organisms such as faecal streptococci or *Clostridium perfringens* should be undertaken, since the presence of these can help resolve the uncertainty. Faecal streptococci occur normally in faeces, but in humans are usually greatly outnumbered by *E. coli*. If organisms in the coliform group, but not *E. coli*, are found in a water sample, the presence of faecal streptococci can afford evidence for their faecal origin. *Cl. perfringens* also occurs normally in faeces, though in much smaller numbers than *E. coli*. The spores of *Cl. perfringens* are capable of surviving in water for a much longer time than vegetative bacteria such as coliform organisms and faecal streptococci. They are also more resistant to chlorination. The isolation of *Cl. perfringens* from water thus suggests that faecal contamination has occurred in the more distant past, and in the absence of coliform organisms and faecal streptococci, that treatment has not been fully effective.

3.4 Other Possible Faecal Indicator Organisms

Coliforms, faecal streptococci and sporulating sulphite-reducing anaerobes are currently recommended for use as indicators of faecal pollution in water. Other micro-organisms that have been proposed for this purpose include the *Bacteroides fragilis* group, *Bifidobacterium* species, *Candida albicans*, *Mycobacterium* species, *Rhodococcus coprophilus* and coliphages.

These groups of micro-organisms are briefly described together with an indication of their value as indicators of faecal pollution. They do not appear to have been evaluated as indicators of such pollution in treated waters. At present they are not considered to be of any value in the routine examination of potable water, but may be useful in certain circumstances. See Appendix G for more details.

3.5 Indicators Distinguishing Human From Animal Faecal Pollution

The ability to distinguish between human and animal faecal pollution may be of value in tracing the source of faecal pollution of a water supply and in assessment of the adequacy of protection of a water supply, especially in rural areas.

In the past, two approaches have been suggested for distinguishing between human and animal faecal pollution. Firstly, an estimation of the ratio between one group of micro-organisms and another (such as between faecal coliforms and faecal streptococci). Secondly, the isolation of a specific indicator such as sorbitol-fermenting bifidobacteria (Appendix G), *Streptococcus bovis*, *Rhodococcus coprophilus* or *Enterococcus faecalis*.

Because of the uncertainly associated with the interpretation of ratio estimates, the faecal coliform : faecal streptococcus ratio is no longer recommended for differentiating human from animal sources of pollution (APHA 1989).

Streptococcus bovis is the predominant faecal streptococcus in the gut of cattle and most farm animals while being relatively uncommon in the human gut (Mara and Oragui 1985). Oragui and Mara (1981) have described a selective medium for the enumeration of this micro-organism, but its usefulness as a specific indicator of animal faecal pollution is limited by its short survival time outside the gut. It could be used when found in conjunction with *Rhodococcus coprophilus*, to indicate the recent nature of faecal pollution from animals; its absence when *Rhodococcus coprophilus* is present would suggest more remote pollution.

Enterococcus faecalis has been used to indicate pollution of human origin as it is particularly associated with man (Mead 1964). However, *Ent. faecium* is considered by others to be more prevalent in human faeces (Guthof 1957, Buttiaux 1958). Care is therefore needed when interpreting results as the prevalence of *Ent. faecalis* depends on the source and its absence does not necessarily exclude human faecal pollution.

3.6 Other Bacteria

3.6.1 Colony counts

Other micro-organisms, such as those associated with soil and vegetation, also occur naturally in ground and surface waters. Many of these organisms are usually able to survive for long periods in the environment and in the warmer months may multiply considerably. Counts of aerobic organisms which grow as colonies on plates of a nutrient agar under defined conditions thus provide a useful means of assessing the performance of water treatment processes. Such colony or plate counts can also provide a general indication of the bacterial content and hence the hygienic quality of water supplies, although the counts themselves have little direct health significance. In practice, changes in the pattern of colony counts of samples from a given supply are usually much more significant than the actual numerical count of any particular sample. A sudden marked change in the colony count of water in a supply may give forewarning of more serious pollution, whereas deviations in the expected seasonal trend may indicate longer-term changes in the supply. Colony counts, if carried out regularly, are also of particular value when water is used for the large-scale preparation of food and drink. Some bacteria can multiply within the distribution network by using nutrients derived from fixtures and fittings or from assimilable or particulate organic carbon in the water itself. See section 4.3.

3.6.2 *Pseudomonas* group

Fluorescent pseudomonads are widespread in the environment and are able to multiply on a wide range of substrates. When these organisms gain access to treated water, they may proliferate in certain circumstances by utilising nutrients either present in the water or derived from unsuitable materials used in the construction of distribution systems or in domestic plumbing installations. They may also grow on membranes during routine coliform testing, though they do not form yellow colonies.

Pseudomonas aeruginosa, an important member of the group, may be derived from human or animal excrement, but its presence in faeces is not universal. Because of this and its ability to multiply in water containing suitable nutrients it cannot be used as an indicator of faecal pollution. Its presence in potable water is, however, undesirable as it can itself act as an opportunist pathogen; its subsequent growth can also be associated with considerable deterioration in microbiological quality of the water. This may affect the colour, turbidity, taste and odour of the water, and result in consumer complaints. Deterioration of this kind is particularly liable to occur, for example, where there is limited flow in part of the distribution system, and also a rise in the temperature of the water. Other fluorescent pseudomonads, especially *Ps. fluorescens* and *Ps. putida*, may also give rise to problems in treated waters by producing slimes during growth which often form the basis of consumer complaints. Since these organisms can grow in so many different situations it is impractical to give any guarantee about their absence from distributed water—however desirable this may be—although they should not normally be present in water for drinking when it leaves the treatment works. In the United Kingdom the enumeration of fluorescent pseudomonads, and *Ps. aeruginosa* in particular, is therefore **not** recommended as a routine procedure, although it may be of value in the investigation of consumer complaints and distribution problems. It may also be of value within certain industries, for example in the manufacture of food, drink or pharmaceutical products, where water of exceptional microbiological purity is often required. In hospitals and other places where immunocompromised persons are particularly prone to infection, *Ps. aeruginosa* may be of some importance as an opportunist pathogen and its presence in water may thus be of concern. In such situations, where water free from *Ps. aeruginosa* is desirable, special treatment facilities may be required on-site. For all these reasons, methods suitable for the detection and enumeration of *Ps. aeruginosa* in water are given in section 7.13, although it is emphasized that monitoring potable supplies for *Ps. aeruginosa* is not recommended as a routine procedure.

3.6.3 *Aeromonas* group

Aeromonas species are natural inhabitants of the fresh water environment and consequently are common in source water. They can infect wounds acquired in water and they are sometimes responsible for septicaemia particularly in patients with an impaired immune response. There is also growing circumstantial evidence that some strains may cause diarrhoeal disease. The study of aeromonads as a cause of infection is hindered by the difficulty in differentiating the genetically definable species by easy practicable chemical tests. In addition, some species appear on membranes during water treatment testing as small, lactose-fermenting colonies which can be easily confused with coliform bacteria. In simple terms it is useful to divide the species into the non-motile group, *A. salmonicida*, and the motile aeromonads. In the context of drinking water, the motile species are of more significance since they include the strains isolated from patients and are common in natural water where they sometimes form a large portion of the total heterotrophic bacterial flora. Motile aeromonads can be present in high numbers in fresh water both in the presence and absence of faecal pollution. High numbers are common in sewage effluent but they are of a type different from those found in pristine fresh water, although all types have been incriminated as possible enteric pathogens. The non-motile aeromonads are pathogens of fresh-water fish but are not as common in water as the motile species.

Aeromonas species are readily killed by chlorine and other commonly used water disinfection processes. However, any survivors of the initial treatment or any aeromonads entering the distribution system downstream through accidental damage, back flow or other causes may grow and multiply significantly. The amount of this growth depends on the availability of nutrients in the water and on the temperature. Thus the presence of *Aeromonas* species in drinking water does not necessarily indicate faecal pollution but reflects inadequacies either in the treatment process or within the distribution system itself. As with pseudomonads, because of their ubiquity there is no need to test routinely for aeromonads. Other determinations such as residual chlorine, temperature and available organic carbon may indicate the potential for aeromonads to grow. However, examination for aeromonads may be useful when investigating distribution problems. Because of their possible pathogenicity their detection in water may be required in pharmaceutical manufacture, food and drink manufacture and hospitals. Accordingly, tentative methods for their isolation are included in section 7.14.

3.6.4 *Mycobacterium* species

The mycobacteria include both obligate parasites and saprophytes. Some species, notably *M. kansasii*, *M. xenopi* and *M. avium*, have been isolated from tapwater and may be pathogenic for humans in certain circumstances (Collins, Grange and Yates 1984). However, the routine examination of potable supplies for these micro-organisms is not recommended because of the prolonged incubation required, the difficulty in identification, and in the interpretation of the significance of any isolate.

3.6.5 Nuisance organisms

These organisms can cause objectionable taste, odour, colour and turbidity in water and may interfere with treatment processes by blocking strainers and filters. They are a morphologically and physiologically diverse group which includes fungi, actinomycetes, iron and sulphur bacteria, algae and protozoa. Most of them are controlled by the usual water treatment processes, but occasionally may establish themselves in sediments, in slimes and on materials within the distribution system (Hutchinson and Ridgway 1977). Routine examination for such organisms is not recommended because of their diverse nature and unpredictable occurrence.

3.6.6 Cyanobacteria (Blue-green algae)

Cyanobacteria are a group of photosynthetic organisms with features of both algae and bacteria. They occur naturally in many inland standing bodies of water, and can often be seen forming a surface scum or "bloom". A bloom is more likely to occur under stable weather conditions, and is brought on more quickly as a result of increased temperature and light. The ability of many cyanobacteria to move through the water and to accumulate at the surface is due to their possession of gas-filled vesicles, which make them buoyant (Walsby 1975).

Problems of taste and odour have been associated with algal blooms. The ability of cyanobacteria to produce toxins is well reported but there is no evidence that these pose any significant risk to health via water supplies.

3.6.7 *Legionella pneumophila*

Legionellas are not normally infective when swallowed, and are therefore not a problem for potable water supplies as such. For this reason details of methods for their isolation are not included in this Report. However, legionellas can colonise domestic and other hot and cold potable water systems, and when aerosols generated from such colonised systems are inhaled infection can result. Attention is therefore drawn to the booklet by the Health and Safety Executive, 'The control of legionellosis including Legionnaires' disease' (HSE 1991) and the BSI methods for sampling (BSI 1992a) and for detection and enumeration (BSI 1992b) of *Legionella* organisms in water, where appropriate guidance can be found.

3.7 Enteric Viruses

The increasing recognition of water-borne enteric viral infections has focused attention on the need to detect viruses in potable water. However, there is little evidence that water suitably treated to comply with bacteriological drinking water standards will transmit viral infection. The Directive requires the absence of pathogens, including enteroviruses, in drinking water and WHO (1993) recommends that drinking water must essentially be free of human enteroviruses to ensure negligible risk of transmitting infection. The Regulations do not mention viruses specifically but require that water supplied for drinking, washing or cooking, or for use in food production, does not contain any organism at a concentration or value which would be detrimental to public health. At present, methods are available only for detecting enteroviruses and rotaviruses. They require large sample volumes, which should be collected as soon as a potential need is perceived, are technically demanding and have a variable efficiency of detection. The routine analysis of potable water for viruses is thus not recommended. The advantages of monitoring for enteroviruses stems from their environmental resistance. As they survive longer in water and are relatively more tolerant of chlorine disinfection than *E. coli* they have the potential to indicate faecal pollution over a longer timescale and at lower levels. Furthermore, methods most generally available are those for isolating certain human enteroviruses. The presence of such viruses in potable water will indicate faecal pollution, though they are not known to be responsible, *per se*, for symptomatic water-borne infection.

Consideration should be given to testing for viruses when there is evidence of post-treatment pollution of a potable water supply. Testing for viruses may also be carried out when there has been a problem with the treatment process, such as a partial failure of disinfection. Examination for viruses may also be done as an additional investigation when there is a suggestion of a possible outbreak of viral infection associated with a potable water supply.

Useful information on the efficacy of the treatment processes may be obtained by monitoring for the presence of enteroviruses in a surface water source and at stages through the treatment works. Increases in the viral content of a surface source water may be an indication for a modification to the level of treatment required.

Details of appropriate methods for detecting viruses in all kinds of water and related materials can be found in 'Methods for the isolation and identification of human enteric viruses from water and associated materials' (SCA 1994). It is likely that advances in technology will, before long, extend the range of available methods to include the detection of more water-borne viral pathogens.

3.8 *Giardia intestinalis, Cryptosporidium* and *Entamoeba histolytica*

Several protozoan parasites, including *Giardia intestinalis* (= *lamblia* = *duodenalis*) and a number of amoebae, can be transmitted in water and may give rise to infections in humans. The life cycle of each parasite is different, and the stage which is detectable in water is the cyst for *Giardia* and the oocyst for *Cryptosporidium*. Provisional methods for the isolation and identification of these and of some amoebae can be found in SCA (1989).

Cryptosporidium species were first recognised more than 80 years ago. They are able to infect a wide range of animal hosts including calves, lambs, piglets, poultry, rodents and dometstic pets, and are now recognised as a significant cause of human gastro-enteritis. Infection is transmitted by means of oocysts, the environmentally hardy stage of the parasite containing the infective stage (DoE/DH 1990).

Small numbers of cryptosporidial oocysts are likely to be present from time to time in most surface waters, particularly after heavy rainfall, and when certain livestock activities and slurry spreading occur. Oocysts can be present in discharges from sewage works and ground waters, which may become contaminated as a result of fractured or damaged sewage systems, and the uncontrolled discharge of septic tanks and cesspools. (See also section 4.1.6).

The cysts of *Entamoeba histolytica* may be found in waters contaminated with sewage. Techniques for the detection of this organism in water supplies have not been properly validated, but it is the general consensus that methods concentrating *Giardia* will be effective for *Entamoeba* species. The detection of cysts morphologically identical to those of *E. histolytica* does not have the same impact as the detection of *Giardia* or *Cryptosporidium* because species of *Entamoeba*, not infective for humans but with morphologically indistinguishable cysts, are found in wild, domestic and farm animals. Currently there is no commercially available antibody stain specific for *E. histolytica* cysts.

3.9 Animalcules in Mains Water

The presence of animals, either whole or in part, has been recognised as a cause for concern. Although their presence may indicate a water supply or water treatment problem, and their detection must always lead to an immediate assessment of microbiological quality, there is no known health risk associated with animalcules commonly found in mains water in the United Kingdom. The source of the water supply and the treatment it receives generally affect the species present, their abundance and their ability to populate the distribution system. A more detailed account of the presence and identification of animalcules in mains water is given in Appendix F.

3.10 Other Aesthetic Problems

An ideal potable water should be clear, colourless, odourless and acceptable to the palate. In practice, the aesthetic properties of a drinking water will depend to a large extent on its source. Water obtained from deep bore-holes in sandstone or chalk will, in its untreated state, generally be nearer the ideal than a water derived from an upland peaty source. The degree of treatment required to render a source water wholesome may also have an effect on its organoleptic properties. In most instances, when a potable water excites adverse comment regarding its appearance, taste or odour, the causes tend to be physical or chemical rather than microbiological. It is nevertheless advisable to examine a sample microbiologically as a check on the general quality of the water. The odour of hydrogen sulphide may result from the bacterial reduction of sulphates. Musty, mouldy or earthy tastes may be the result of growth of fungi or actinomycetes in the cold water pipes in buildings where the temperature of the cold supply is undesirably high (Holden 1970). Increased chlorination following, for example, repairs on the distribution system is sometimes responsible for complaints regarding taste.

Chapter 4: Maintenance of Water Quality

4.1 Source Protection

4.1.1 Introduction

No source of water that is intended for human consumption can be assumed to be free from pollution. All sources have different microbiological qualities and may be subject to natural or man-made sources of pollution which may result in the deterioration of water quality to the point where treatment is no longer effective in removing the contamination. If this happens it may be necessary to use an alternative supply until the pollution is remedied.

Much can be done to limit pollution by regular inspection and monitoring of water sources, by encouraging the use of good agricultural practice and by controlling effluent discharge. Guidance on these matters is given in the Water Authorities' Association report (WAA 1988), the Code of Good Agricultural Practice (Ministry of Agriculture Fisheries and Food and Welsh Office Agriculture Department) (MAFF/WOAD 1985), and Policy and Practice for the Protection of Groundwater (National Rivers Authority) (NRA, 1992). This guidance should be followed by all those responsible for the protection of water sources.

4.1.2 Sources of pollution

4.1.2.1 Natural pollution

The quality of many source waters will depend upon geology, soil type, natural vegetation, climate and run-off characteristics. Disruption of natural geology and heavy rainfall can dramatically affect water quality. Wild animals and birds can also be natural sources of human pathogens, see section 2.1.1.

4.1.2.2 Agricultural pollution

All types of water sources may be subjected to contamination by agricultural activity. Free-range animals may deposit excreta into water and animals like cattle also have a habit of wading and may stir up sediments. Much can be done to reduce the risk of water contamination from slurry spillage, or the use of slurry on land followed by surface run-off, by considering controlling access to land, reaching agreements with land owners over agricultural practices and adopting aquifer protection policies.

4.1.2.3 Recreational activities

Recreational activity may cause pollution through direct contamination of water with faecal material or indirectly by faulty drainage or leakage from sewers and septic tanks provided as part of public access facilities. Proper control of recreational activities or treatment commensurate with the recreational use of water should provide adequate protection. Where the public has access to reservoirs, consideration should be given to the provision of toilets and handwashing facilities and where these are provided they should be well maintained and inspected regularly.

4.1.2.4 Effluent and waste disposal

The discharge of effluents from sewage treatment works, septic tanks and cesspools can dramatically increase the microbial content of surface waters. The installation of septic tanks and cesspools should be in accordance with the British Standard Code of Practice for the Design and Installation of Small Sewage Treatment Works and Cesspools, BS6297 (BSI 1983). The discharge of industrial effluents, particularly from abattoirs and cattle markets, may also contain large numbers of pathogenic micro-organisms which increase the risk of contamination. Slurries and solid material from sewage treatment and animal waste should be spread on land only with strict control, and water undertakers are recommended to observe the Code of Practice for the Agricultural Use of Sewage Sludge (DoE 1989) in this matter. Due notice must be taken of any zones of protection and every effort made to prevent accidental spillage or heavy surface run-off following rainfall.

4.1.3 Water sources

4.1.3.1 Upland reservoirs

The microbiological quality of upland reservoirs, which depend for their supply on catchment areas and feeder streams, is usually very good. Plans should be available showing any installation which may cause pollution and such plans should be kept up to date. All catchment areas should be inspected regularly and agricultural and recreational activities associated with reservoirs should be controlled to minimise excessive contamination by spillage, drainage or access of animals and people.

4.1.3.2 Rivers, lakes, lowland and pumped storage reservoirs

The microbiological quality of these source waters may be much poorer than that of upland reservoir waters necessitating adequate treatment regimes to be developed. Occasionally, quality may deteriorate suddenly because of the spillage of effluents, or heavy rain causing direct run-off from land or the operation of storm overflows. Where such incidents occur, treatment may no longer be wholly adequate to protect the water supply. Water abstracters should ensure that they are aware of potential sources of pollution into rivers, and any pollution incident likely to affect water treatment should be notified to them as quickly as possible. The provision of bankside storage greatly reduces the impact of pollution events.

4.1.3.3 Springs, wells and boreholes

Groundwater forms an important part of water resources and provides up to 35% of the potable water supplies in the United Kingdom. It is usually of much better quality than surface water and often only requires chlorination before being put into supply. Aquifer contamination may be caused directly by the seepage of material or by the movement of groundwater of poor quality into areas of better water quality. Because aquifer pollution can exist for very long periods of time and may be impossible to remedy, it is preferable to take preventative measures to minimise pollution rather than to provide remedial action or develop alternative sources of water, which may be expensive.

Spring water should be collected in secure underground chambers and covers should be raised clear of the ground and surrounding vegetation. Wells and boreholes should be lined to a depth sufficient to prevent the entry of any surface pollution or polluted sub-soil water. There should be an effective seal between the lining and the ground. The head of a well or borehole should be protected and the installation should be checked at regular intervals to ensure that ingress does not occur. Up to date records of the percolation area must be kept showing any sewers, septic tanks, cesspools, waste disposal sites, soakaways or any other sources of potential pollution. Water undertakers should be consulted if any new sites or structures are proposed. There may be areas where the establishment of a protection zone might be considered to prevent contamination of the aquifer. Consideration should be given to the geological nature of the aquifer and the presence or absence of any overlying drift which may provide additional protection. The accumulation and spreading of human and agricultural wastes might be prohibited or restricted in protection zones.

4.1.4 Monitoring source waters

Regular monitoring of source water is essential in order to provide information about general water quality and to establish seasonal variations in water quality and changes in response to weather and other factors. Under circumstances where the microbiological quality of the water deteriorates or where there is suspicion of microbial contamination, water abstractors should initiate immediate investigations and take appropriate remedial action. Consideration should be given to the introduction of more detailed microbiological monitoring including additional tests for bacteria, viruses and intestinal parasites. New water sources of which the microbiological quality is unknown should be subjected to intensive microbiological monitoring to establish the water quality and to ensure that proposed treatment regimes are adequate. The Regulations include specific requirements to monitor new sources.

4.1.5 Action following a pollution incident

It is essential for water undertakers to have plans to cope with pollution incidents and these should be rehearsed regularly. A comprehensive list of channels of communication should be available at all times and this should be kept up to date. Laboratory facilities for microbiological analysis should be adequately staffed and equipped to cope with emergencies.

4.1.6 *Cryptosporidium*

Following an outbreak of cryptosporidiosis in Swindon and Oxfordshire in 1989 a Government appointed Expert Group, under the chairmanship of Sir John Badenoch, was convened to advise upon the significance of *Cryptosporidium* in water supplies.

The Expert Group reported in 1990 (DoE/DH 1990) and its recommendations resulted in the establishment of a national collaborative programme of research which has been steered by the Drinking Water Inspectorate. The programme included the following studies: occurrence and distribution; removal and inactivation in water treatment; methods of anlysis; assessment of viability; and infectious dose. The Report of the Expert Group should be consulted for recommendations on means of reducing the risk of infection, preventing oocysts of *Cryptosporidium* from gaining access to water supplies, monitoring strategies, and investigation and management of outbreaks.

In July 1992, the Drinking Water Inspectorate published an interim report on the national research programme which included a review of progress worldwide (DWI 1992). It was apparent that there was a need for further work on more precise quantification of the level of risk and establishment of the degree of catchment protection and type of water treatment necessary to eliminate the risk, or to reduce the risk to an acceptable level. This programme of working is now drawing to a conclusion and the results will be published through thc Water Research Centre and the Foundation for Water Research.

4.2 Water Treatment

4.2.1 Objectives of water treatment

The objective of water treatment is to produce a final water which is microbiologically and chemically safe for consumption, is not corrosive towards materials in contact with water and is aesthetically acceptable. The range of treatment processes includes clarification and sedimentation, filtration and disinfection. Depending on the source and nature of the water, one or more of these processes can be used, each further preparing the water physically and chemically for disinfection. Treatment to make water microbiologically safe to drink constitutes a complex and highly technical field and for further information the following should be consulted: Cox (1964), Skeat (1969), Holden (1970), Hutchinson and Ridgway (1977), the Standing Committee of Analysts (SCA 1980), White (1985), and Hall and Hyde (1992). Whilst each of the treatment processes is able to reduce the numbers of particular micro-organisms, no process can ever ensure their complete removal; in the UK, therefore, disinfection (usually by chlorination, although other techniques are becoming more popular) is the final essential safeguard against water-borne microbial disease.

When chlorine is used the dose should be so selected that the chemical demand of the water is satisfied and that an adequate contact time is achieved before water is distributed to any consumers (WHO 1993). An appropriate chlorine residual should be maintained throughout the distribution system and should preferably be present in water at consumers' taps, to reduce the risk of microbial regrowth and to provide an indication of the absence of post-treatment contamination (WHO 1993). It is essential therefore that the chlorine residuals are monitored regularly both at the works and in distribution. It is important to distinguish between the different forms of chlorine (free or combined) and to check the pH and temperature of the water. These and many other factors have a marked effect on the efficacy of chlorination (SCA 1980).

Although micro-organisms differ in their susceptibility to chlorine (in decreasing order of resistance: protozoan cysts, bacterial spores, enteric viruses and enteric bacteria) the combination of chlorine concentration and contact time necessary for inactivation of intestinal viruses and pathogenic bacteria can be readily achieved by a properly designed and operated treatment works. It should be noted that certain incidents of water-borne disease in the United Kingdom and elsewhere have occurred as a result of inadequate chlorination or because no such facility was installed or in use (PHLS 1978, Galbraith et al 1987, Benton et al 1989).

Attention has also been drawn recently to the possible health effects of certain chemical compounds formed as by-products of chlorination. The World Health Organization (WHO) International Agency for Research on Cancer (IARC) concluded in 1991 that "there is inadequate evidence for the carcinogenicity of chlorinated drinking water in humans" (IARC 1991). Expert advice was obtained by the Department of Health in April 1992. The committee on Carcinogenicity of Chemicals in Food, Consumer Products and the Environment concluded that modification of chlorination processes or their replacement was not required on the available data on cancer epidemiology, animal carcinogenicity and mutagenicity in relation to chlorination by-products in drinking water (DH 1993b). It is prudent to reduce the concentration of disinfection by-products if this can be achieved without prejudicing the microbiological quality, and to continue to identify by-products and assess their potential for affecting health. It also continues to be the case that the microbiological safety of potable water supplies is of paramount importance (WHO 1993) and the benefits of chlorination are re-affirmed. However, it is also considered important to re-emphasise that when sources of surface water, or groundwater impacted on by surface water, are utilised then the installation of multiple barriers to the transmission of infection is strongly recommended as a series of safeguards against water-borne disease (WHO 1993).

While the proper design, operation and maintenance of treatment works is of the utmost importance, frequent microbiological monitoring is necessary for adequate assessment of the hygienic quality and safety of drinking water. Chemical and biological tests—apart from those required for treatment control purposes—can be made less frequently. The information derived from microbiological tests must however be assessed in the light of thorough practical and working knowledge of the conditions applying at the source, throughout all the stages of treatment to which the water is subjected, and in the distribution system itself.

It is particularly important that the microbiologist should consider the implications of the sudden pollution of a previously satisfactory supply Failure or inadequacy of treatment processes, particularly chlorination, can be very serious, but other hazards may cause deterioration in microbiological quality. These include contamination via air-valves and stop valves, infiltration into mains and service reservoirs, cross-connections, back flow and venturi effects. Sudden changes in the microbiological quality of ground waters can occur through cesspool leakage, from accidental or illicit contamination of the gathering grounds or by polluting material gaining access through faults or fissures in the strata or through defects in the well or borehole lining. Heavy rains following prolonged drought may enhance the risk of pollution of a water source and service reservoirs if the structure is unsound. Increased pumping from wells, perhaps as a result of prolonged drought, may also lead to the pollution of previously satisfactory sources. Whenever these or other environmental conditions occur the frequency of microbiological examination should be increased; the points of sampling being carefully chosen so that any changes in quality may be identified quickly and appropriate action taken.

4.2.2 Point-of-use and point-of-entry devices

Public water supplies drawn directly from the mains do not on health grounds require further treatment. In some instances, however, consumers may object to the taste or appearance of a mains water supply, or to certain of its constituents, and may wish to make use of a device to remedy these. This is acceptable provided that the device and its use comply with the water supply bye-laws. The device may be fitted to the supply as it enters the premises (point-of-entry device) or at a single tap (point-of-use device), usually in the kitchen. Such a device may also provide a convenient means of improving the quality of a small private supply. If it is used to ensure that a private supply complies with the legislative requirements it will be necessary to site it at the point of entry to the premises and it will need to be capable of fully treating the water. Details of treatment devices suitable for private water supplies are given in the Manual of Treatment for Private Water Supplies (DoE/WO/SO 1993).

4.2.2.1 Disadvantages of point-of-use and point-of-entry devices

Devices that incorporate some form of filter medium or ion-exchange resin will often allow bacteria to proliferate within the unit, so that in time, water leaving the device may yield increasing bacterial counts. Activated carbon filters in a warm environment provide an ideal growth medium for many bacteria and regular maintenance, especially regular renewal of the unit, is essential if the quality of the water leaving the filter is to remain acceptable (Geldreich et al 1985).

The use of these devices may also initiate adverse health effects. Ion-exchange resins may remove some desirable elements such as calcium from consumers' diet while simultaneously increasing the intake of sodium. Bacterial colonisation of reverse osmosis units has been linked to gastro-intestinal symptoms (Payment et al 1991), although the causal mechanisms remain to be explained.

4.3 Biofilms

In low-nutrient aqueous environments, micro-organisms preferentially colonise surfaces rather than grow in the planktonic phase. This gives them nutritional advantages and can protect them from adverse environmental influences. The organisms colonising surfaces become embedded in a matrix of extracellular polymeric substances which are excreted by the organisms themselves. This layer of growth is termed a biofilm and in water distribution systems is usually quite thin, not exceeding a few hundred microns. An appreciation of the properties and role of biofilms is imperative for an understanding of the microbiology of man-made water systems. In both the natural environment and water distribution systems biofilms are usually composed of complex consortia of micro-organisms including bacteria, fungi and protozoa. The metabolic by-products of one organism can provide nutrients for others. This enables organisms, that would otherwise be unable to grow by themselves, such as *Legionella pneumophila*, to proliferate. Biofilm distribution is patchy and can vary considerably even over distances of a few millimetres or less. Biofilms can also accumulate organic and inorganic debris from external sources by the adsorption of silt, sediments, inorganic precipitates and corrosion products. These materials can provide extra nutrients for microbial growth.

No material that comes in contact with water is immune to colonisation but some materials may support or promote more growth than others. To maintain water quality during distribution, construction materials should not promote growth. Non-metallic materials should comply with BS 6920, which includes a test for growth promotion. An equivalent standard does not exist for metals which, although they can become colonised with biofilms, cannot provide organic nutrients for growth.

Most of the growth that takes place in distribution systems probably occurs in biofilms and the majority of the planktonic organisms are derived from organisms leaving the biofilm or by the biofilm sloughing off. Organisms that have survived disinfection, including coliforms, can become attached to biofilms where they may subsequently grow. Biofilms are important because they contribute to many of the problems that can occur in water distribution systems. They may promote or cause corrosion of pipes, can be responsible for off-flavours, contribute to discoloured water, harbour pathogens such as *Legionella* species, increase the chlorine demand and provide a site for the regrowth of coliforms. Biofilms also protect organisms from disinfection. The contact time required to produce a particular degree of disinfection of organisms in the biofilm by chlorine may be hundreds or even thousands of times greater than that required to get an equivalent degree of death for the same organisms suspended in water (LeChevallier, Cawthon and Lee 1988). Thus it is possible for biofilms to continue to survive and grow even when the water contains residual chlorine at the concentrations used in drinking water. This reduction in disinfection efficiency is caused by the diffusion of the disinfectant being reduced by the biofilm and is less for monochloramine than chlorine. From the foregoing discussion it is clear that the efficient disinfection of a water system depends on the control of biofilms and disinfection failures may occur through failure to take this into consideration.

Measuring the degree of microbial contamination of water (planktonic phase) gives a poor estimate of the total microbial activity in a water system. Unfortunately it is difficult to measure the degree of biofilm formation routinely because of problems of collecting representative samples of biofilm. Other physical and chemical determinations such as total and assimilable organic carbon, dissolved oxygen and temperature could be made in areas where biofilms are considered to be a problem. It should always be remembered that the inability to detect micro-organisms in water samples does not mean biofilms are absent.

4.4 Microbiological Monitoring of Drinking Water

4.4.1 Routine monitoring—general considerations

The Regulations require water undertakers to define water supply zones in which not more than 50,000 people normally reside. Samples are required to be taken from consumers' taps in these zones at specified frequencies related to the number of people residing in the zone. Samples are required to be taken of the water leaving treatment works at frequencies related to the volume of water or population supplied. Water in all service reservoirs including water towers is required to be sampled once in each week the service reservoir or water tower is in use.

For each supply zone, the number and location of sampling points must be determined to ensure, as far as possible, that the data produced are representative of the quality of the water supplied in that zone. At least 50% of the sampling points for microbiological and residual disinfectant analyses must be selected at random.

"The Private Supplies Regulations" also specify numerical standards for identical microbiological parameters. It is the duty of local authorities to take and analyse samples of water according to prescribed frequencies depending upon the classification of the private water supply.

4.4.2 Action when standards are infringed

Immediate action is essential whenever there is any indication that a microbiological standard has been infringed. At the least, this will require the examination of another sample taken from the same and/or a closely related sampling point together with confirmation of the nature of the indicator organism found in the original sample. Further action may be required and details are given in Chapter 7 of the Guidance Document and in sections 2.1.2 and 4.6.2.2.

4.4.3 Reviews of the quality of a water supply

The microbiological quality of water supplies should be reviewed regularly. This may provide evidence of seasonal variations in quality in a particular zone which may not otherwise be apparent. Records of the results of all laboratory tests should therefore be maintained to enable comparisons to be made. In addition, the Regulations require water undertakers to provide local authorities, at least annually, with information on the quality of the supplies in their area.

Total coliforms should always be absent from treated water in supply. In practice, this ideal may not be consistently attainable and is recognised in the Regulations. Water shall be regarded as satisfactory in respect of the requirement as long as 95% of all samples taken during the previous year (or the last 50 samples, if fewer than 50 have been taken during the year) have not contained total coliforms. The provision does not, of course, apply to faecal coliforms (*E. coli*). Any intermittent and related low-level failures will require investigation and remedial action.

Regular colony counts provide an indication of overall microbiological quality and each supply zone will have its own particular 'normal' range. Any significant change in this range, suddenly or as a trend over time, and especially of counts at 37°C, will require investigation and, where necessary, remedial action. For colony counts, the Regulations prescribe there should be no significant increase in levels over those normally observed.

4.5 Microbiological Sampling for Operational Evaluation

4.5.1 Raw water sampling for process evaluation

The Regulations require monitoring of a new source before it is put into supply to establish whether it is wholesome and what treatment is necessary to render it wholesome. In situations when raw water is prone to pollution there should be contingency plans for providing additional treatment, particularly chlorination, or to provide alternative supplies. In such circumstances there may be a need to consider automatic continuous monitoring of water quality and residual disinfection control systems. In any case a more intensive investigation should be undertaken when any unexplained deterioration in quality occurs.

4.5.1.1 Frequency of sampling

The Water Authorities' Association Report (WAA 1984) gives detailed information and frequencies for raw water monitoring, including comprehensive consideration of microbiological monitoring. Upland reservoirs will show seasonal variations in water quality. Catchment control is essential and guidance is given in the Water Authorities' Association Report (WAA 1988).

Underground fresh water supplies are expected to be free of pathogens. Experience has shown however that groundwater can become microbiologically contaminated (Short 1988) so regular examinations for bacterial indicators and colony counts are recommended for all borehole sources.

Samples of water should always be taken before boreholes, which have been pumped to waste or left standing, are put back into use. There may also be a need, in circumstances where pollution has occurred, to examine the water for bacterial, viral and protozoal pathogens.

4.6 Investigational Sampling

4.6.1 Water at the point of use

Compliance monitoring programmes based on periodic sampling do not provide absolute or continuous assurance of attainment of a satisfactory microbiological quality at all points where water is available for use by consumers. For this reason undertakers may receive valuable feedback from their consumers, indicating where a local deterioration in water quality may be occurring. Effective monitoring of consumer communications is important and can assist in the investigation of microbiological quality problems.

Many local authorities and water undertakers collect investigational samples for microbiological examination whenever they become aware of complaints of ill health that could possibly be related to drinking water. Many undertakers also collect microbiological samples in the course of investigations into aesthetic or other quality complaints. These investigations lead to considerable knowledge about water quality at the point of use and may reveal incidents of back-flow or other causes of water quality deterioration in distribution and plumbing systems. Formal procedures for investigating, recording and reporting all consumer complaints, will assist management of these complaints.

An immediate response should be made to all consumer complaints, especially if there is any suggestion of ill-health that may be related to drinking water. Sampling for coliforms, including *E. coli*, and where necessary, additional microbiological indicators such as faecal streptococci should ideally be carried out on the same day as receipt of such a complaint, together with any other chemical, biological and physical parameters deemed appropriate.

It is essential that those who deal with complaints from consumers are issued with clear guidelines describing the information to be obtained from complainants. The practice of collecting a microbiological sample wherever feasible in response to complaints concerning the quality of water supplies is commended as a means of providing further water quality data.

4.6.2 Sampling strategy for investigative purposes

4.6.2.1 Response to consumer complaints in building premises

Whilst routine sampling assesses the overall quality of the water being supplied to consumers, investigational samples are aimed at establishing the quality of a particular supply at the point inside the premises where it is available to the consumer. It is important therefore to identify and sample the tap or fitting used for potable purposes. In some instances the tap will not be connected directly to the rising main but served via a storage cistern or other fitting. In these situations reference samples should be collected from a tap connected to the rising main, sited as close to the incoming point of supply as possible. If there is no such tap, or the circumstances suggest that contamination could be arising from within consumers' installations, then samples should be collected from a standpipe set up at the closest hydrant point upstream of the premises or from a tap in a neighbouring property on the upstream side. The use of these sample points will facilitate rapid interpretation of the analytical results and site survey findings. The results should be reported to the appropriate local authority and district health authority. Where the cause of water quality deterioration is established, agreement should be reached on the most appropriate remedial measures. Where the problem arises within consumers' installations then the owner or appointed representative should be advised on how to remedy the problem and where appropriate, be advised of the consumer's liability under the water bye-laws.

In large premises, considerable detail may be required in order to determine the number and location of appropriate sampling points. Ashworth and Colbourne (1987) report a useful checklist which summarises the procedure for examining a wide range of quality problems in buildings. A full inspection of the installation for compliance with water supply bye-laws is useful when there is information to suggest a quality problem is linked in time to modification or extension of plumbing systems.

4.6.2.2 Response to routine sample 'failure' in distribution systems

When coliform bacteria are reported in any treated water sample it is necessary to investigate whether the result is indicative of a deterioration in the water in supply or merely an artifact of the sample or sampling point. Results of samples collected at the same time from other points within the distribution zone and associated treatment works should be examined. This may yield useful corroborating information, particularly if the samples have been collected as part of a systematic, as opposed to a random, sampling programme. Other information, including historical water quality data, reports on the physical condition of associated service reservoirs, operational data on the supplying works, remedial work on mains and prevailing weather conditions should also be sought and examined.

When coliform bacteria are detected in a sample, investigative samples should be collected as soon as possible. The investigative samples should include a repeat sample from the same sample point and where appropriate, additional associated samples. In the case of a presumptive coliform isolated from a service reservoir, additional samples may include samples from both the inlet and the outlet mains and from the distribution system fed by the reservoir. The distribution samples should preferably be collected from consumers' taps. If the presumptive coliform is isolated from a sample taken from a property or hydrant, repeat samples should be collected from the tap originally sampled, together with samples from consumers' taps at locations immediately upstream and downstream of the original sampling point.

4.6.2.3 Approach when faecal contamination is suspected

The immediate aim of the investigation is to establish whether the water in supply is contaminated with organisms indicative of faecal contamination. If the original unsatisfactory result indicates the presence of coliforms in the absence of *E. coli* then the investigative samples should be examined not only for coliforms and *E. coli* but also for additional indicator organisms such as faecal streptococci. If *E. coli* was present in the original sample then consideration should be given to the need to collect and preserve samples for examination for pathogens, particularly those which may be unaffected by the presence in the water of residual chlorine e.g. enteroviruses or protozoal cysts. Analysis of these samples need only to be performed if the investigational sampling and analysis confirm that the supply was subject to faecal contamination. The recommendation to obtain such samples at this early stage is based on experience from those instances where water supplies have been linked epidemiologically with water-borne disease. If such sampling is initiated only after notification of cases of disease, then the time delay makes it extremely difficult to obtain microbiological evidence to support the epidemiological findings.

4.6.2.4 Approach when faecal contamination has been ruled out

It is not uncommon for coliform organisms to be isolated from treated waters in the absence of *E. coli*. Environmental strains of coliforms are able to colonise the surfaces of pipes and survive in deposits in reservoirs where they will be protected from residual chlorine or survive other hostile conditions; however, their presence is undesirable given their primary role as indicators of the likelihood of more serious contamination. For this reason the site of colonisation and the source of nutrient must be sought by the water undertaker and priority given to implementing measures necessary to control the growth of these organisms. Investigative techniques that are useful for resolving such incidents include the collection and analysis of larger volume samples (1 litre), deposits and biofilms, the evaluation of construction materials by methods described in BS 6920 Part 2.4 (BSI 1988) and identification of isolates to species level.

Distribution systems are monitored at varying frequencies for counts of bacteria capable of growth at 37°C and 22°C. Where counts are considerably higher than previously reported, an investigation should be initiated. In the first instance this should comprise a review of bacterial counts from associated sample points and consumers' taps within the previous month. If this indicates that a sustained rise in bacterial counts has taken place in all or part of the distribution zone then arrangements should be made for colony counts to be performed on all microbiological samples collected from the affected area and its associated reservoirs and source works. In the absence of evidence of faecal contamination or a failure in treatment at the supplying works then elevated colony counts should not be viewed with concern in terms of the health of the population as a whole. However, the increased likelihood of aesthetic problems (for example, taste and odour) should be appreciated.

If elevated counts persist for longer than one month then an attempt should be made to characterise the predominant bacterial species, as these may be opportunist pathogens such as *Aeromonas* species and *Pseudomonas aeruginosa*. It is possible that other opportunist pathogens such as legionellas and mycobacteria may also be present but not detected, due to their inability to grow in the colony count medium. All these organisms may be present in low numbers in mains and reservoirs and are usually associated with pipe surfaces and deposits. Their appearance in large numbers in the water indicates that conditions in the distribution system have become suitable for growth as opposed to survival of these organisms. This may be as a result of changes in nutrient loading of the source water, an influx of material into the system, a change in environmental conditions (for example, a rise in temperature) or localised stimuli such as corrosion or use of an unsuitable construction material. If it can be established that the elevated colony counts are a persistent and not a transient problem in a distribution zone then attempts should be made to identify the cause and formulate appropriate remedial measures. Additional analysis for other chemical, physical and biological parameters is an important part of any such investigation.

4.6.3 Response to routine sample 'failure' in treatment works

The detection of any coliform organism or a significant increase in the colony count in a sample of fully treated water from a pumping station or treatment works requires immediate investigation. In addition to collecting a repeat sample of treated water, consideration should be given to the collection of water, at the same time, from water at each stage in the treatment process and the raw water. It is especially important to collect a sample of water from taps connected directly to points just before and after addition of chlorine to the water (contact tank).

Investigative samples from all these points should be collected daily and at different times on each day until it has been demonstrated that coliforms are not present in the water passing into supply. Where coliforms are isolated from water entering the distribution system, distribution samples should be taken to determine the extent of any disseminated contamination. The performance of the plant must always be regarded as inadequate until it can be demonstrated unequivocally that either; (i) the sampling facility is not representative of the water in supply; (ii) the sample tap does not comply with the requirements given in section 6.4.6.1; (iii) the samples were in a batch which failed laboratory quality control procedures; or (iv) the source of the contamination of the water or fault with the treatment process has been remedied. Whilst water free from coliforms may be achieved by immediate remedial measures such as increased dosing with chlorine or removal of a given source in the case of (iv), the efficacy of the water treatment processes should always be re-evaluated. This can be achieved by the collection and analysis of large volume samples of both raw and treated water which should be examined for secondary bacterial indicators such as faecal streptococci and other more resistant organisms such as enteroviruses, *Giardia* and *Cryptosporidium*.

These procedures are outlined in Figure 1, which is not intended to be exhaustive.

Where any failure is regarded as giving rise to, or is likely to give rise to a significant risk to health of persons residing in the area, the Drinking Water Inspectorate, Local Authority and District Health Authority must be informed. Prior consultations between Local Authorities, District Health Authorities and water undertakers may be necessary to establish what constitutes a significant risk.

Figure 1 Outline of action following a sample failure

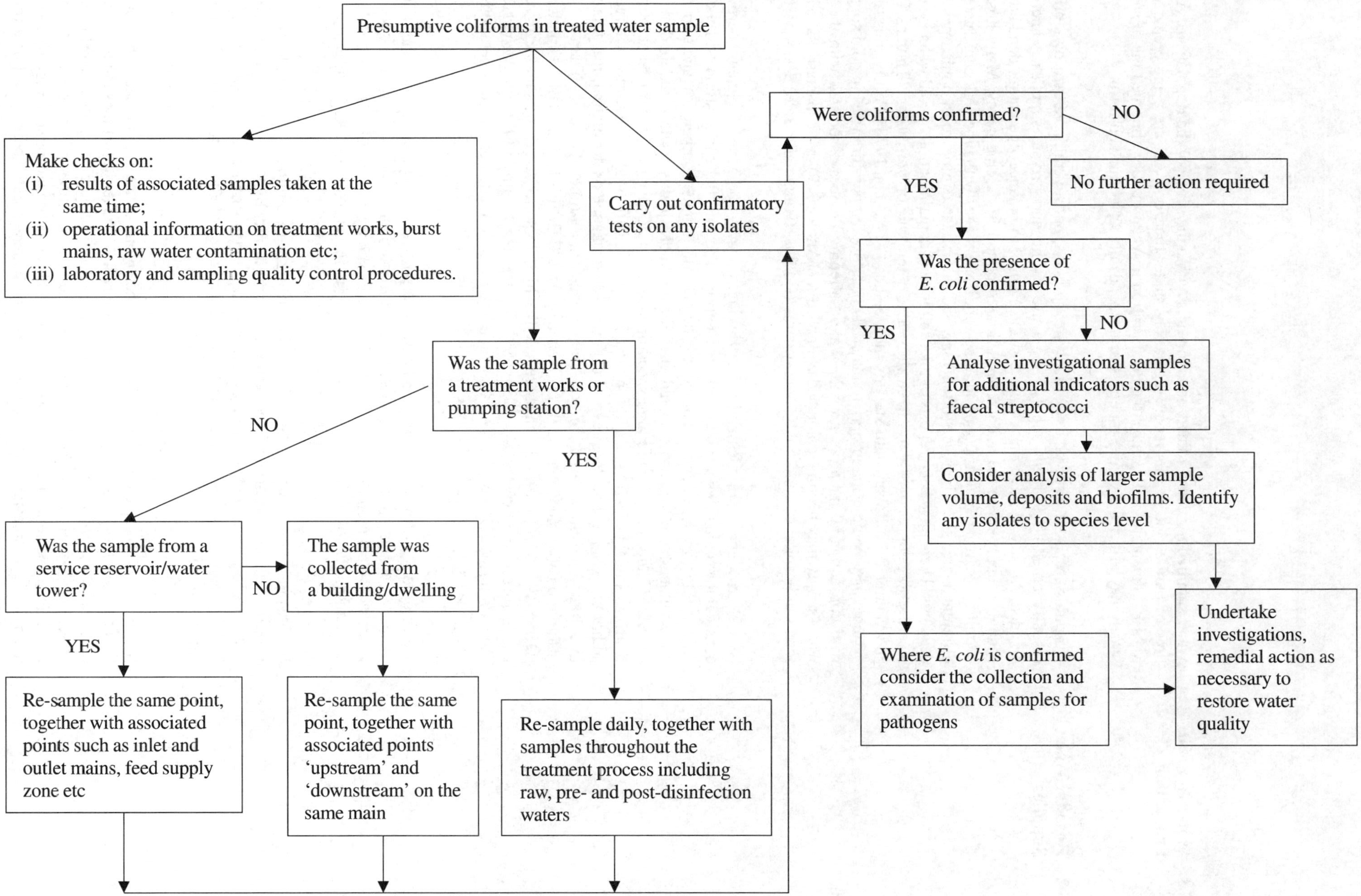

Chapter 5: Special Considerations

5.1 Water in Food Production

Water supplied for use in food production falls within the scope of the Regulations. At the time of publication of this report regulations under the Food Act are being prepared by MAFF. Legislation will control the deterioration in water quality after the time of supply which may affect the wholesomeness of the foodstuff in its finished form.

5.2 Hospital Water Supplies

Hospitals are usually served by public water supplies, but several have their own private (usually borehole-derived) supply. Some use public and private supplies, sometimes blended. Most will have one or more storage tanks to balance fluctuating demand and enable a supply to be maintained for a limited period in an emergency. Many hospital complexes consist of a variety of buildings of different ages with pipework constructed of different materials. There are often long and complicated pipe runs, sometimes with deadlegs. There have been reports of microbiological contamination of cold water systems where storage tank covers have been allowed to deteriorate or were absent (Thomas, Piper and Maurer 1972, Tabaqchali, Stevens and Gazides 1977, Anon 1978). Poorly constructed systems may contain inappropriate materials or allow temperature rises in cold supplies with consequential bacterial growth. Water softening devices associated with the cold water supply may become colonised with bacteria (Anon 1979). Mycobacteria have been found in hospital hot and cold water systems (Collins, Grange and Yates 1984).

As well as water of potable quality for drinking, washing and general hygienic purposes, considerable quantities are required for catering. The hospital's population will include patients of extremes of age and of varying degrees of immune competence, and so may be more vulnerable to water-associated infections. Potable quality water is required for pre-operative hand washing and high speed drills used in dentistry. Water for renal dialysis, pharmaceutical preparations and for some purposes in pathology laboratories will require additional treatment such as reverse osmosis, ion-exchange or distillation. It is therefore important to ensure a supply of good quality water at all times. General guidance on the installation of a cold water supply is given in the DHSS Health Technical Memorandum 27 (DHSS/WO 1976, currently under revision) which contains details of suitable filtration and disinfection processes. The Code of Practice on The Control of Legionellae in Health Care Premises (DH/WO 1988) also includes a section on the design, operation and maintenance of hot and cold water services systems.

Water used for drinking, washing and culinary purposes must comply with the requirements of the relevant regulations whether derived from public or private supplies. Routine microbiological monitoring of a hospital's private supply will therefore be required by its local authority for the parameters specified in the Private Supplies Regulations.

For hospitals with their own private supply, the regular examination of samples at source is recommended in order to detect any variation in quality that may require an adjustment to any treatment carried out. It is also strongly recommended that, regardless of the type of supply, regular checks of the microbiological quality of the water during distribution should be made.

If a sample is found to contain coliforms, or if the colony count is significantly higher than the usual level, immediate investigation is required. At the least, further samples from the same and other points served by that part of the supply should be examined. The hospital's Infection Control Doctor (ICD) must be notified immediately of any such occurrence. Close and continuous liaison between the hospital estates department and the ICD should in any case be an established feature of hospital management. There should be a routine maintenance programme including regular inspection of taps, shower heads, storage tanks and other installed devices.

If major problems arise, the CCDC should be advised and assistance sought from the local environmental health department and the nearest public health laboratory. The CDSC may also provide assistance. If there is major contamination, it may be necessary to boil the water to be used for drinking, food preparation etc or to obtain an emergency tanker or bowser supply. The use of bowsers should be carefully supervised to ensure that the contents do not become contaminated—in any event the water quality should not be taken for granted and all bowser water used for drinking or cooking should be boiled before use.

5.3 Trains, Aircraft, Ships, Bowsers etc

Where water of potable quality is required in circumstances that prevent it from being supplied directly from a mains piped supply, special care will be required to ensure that it is wholesome and remains so throughout the duration of the supply. This applies particularly to potable water supplies in trains, long-distance coaches, aircraft and ships where the supply may have to be maintained over considerable periods from storage tanks.

All storage tanks, bowsers etc that are used for potable supplies must not be used for any other purpose and must be so constructed that they can be completely drained and emptied. Materials of construction should not be capable of supporting microbiological growth, and must be capable of withstanding chlorination to a level of 50 mgL^{-1}. The connecting pipework should be similarly constructed. Bowsers and tankers should be fitted with lockable lids and taps to prevent contamination. All such water storage facilities must be cleaned and disinfected by chlorination before being filled with water of potable quality. They should be regularly drained, cleaned and disinfected at appropriate intervals. Care must be taken during the filling process to ensure that contamination, for example of nozzles, does not occur. Staff engaged in handling these water supplies should maintain a high standard of personal hygiene and not be required to perform other duties that might involve the handling of contaminated materials (such as from toilets etc). They should also be subject to regular health surveillance. Tanks should be drained and cleaned if necessary after each period of use.

The microbiological quality of water supplied by such means should comply with the standards in the Regulations. 'In use' samples should be tested at appropriate intervals to ensure that quality is maintained. As the necessary measures taken to ensure the delivery of water of potable quality will depend upon the nature of the vehicle concerned, further guidance is detailed below.

5.3.1 Trains

Storage tanks on railway buffet cars should be filled with water from a piped main via a standpipe and hose, or from a bowser filled from a standpipe. The standpipes and hoses should be protected from contamination and flushed before filling the bowser or storage tank. The tanks and bowsers should be drained, cleaned if necessary and disinfected at regular frequent intervals. Hose nozzles should be kept immersed in a suitable disinfectant solution when not in use. Where an ultra-violet treatment unit is fitted in the pipe system between the storage tank and the point of use, it must be regularly maintained. Details of the necessary procedures are given in the report of a joint British Rail/Institution of Environmental Health Officers working party on the supply of water for train catering (BR/IEHO 1985). Appropriate procedures should also be applicable to the supply of wholesome water on long-distance coaches.

5.3.2 Aircraft

In general, supply arrangements are similar to those outlined for trains. However, the source water may not always be of potable quality and it may be necessary to chlorinate the water before use. Hypochlorite solution or a solid chlorine-donor sufficient to maintain a residual chlorine concentration of 0.2 mgL^{-1} should be used. More detailed information can be found in the 'Guide to Hygiene and Sanitation in Aviation' (WHO 1977).

5.3.3 Ships

The supply system must be used exclusively for the distribution of potable water. Care must be taken to maintain the integrity of the system and to avoid the possibility of contamination. Cross-connections with non-potable water systems and back flow must be avoided. Water taken on board should be of potable quality and care must be taken to ensure that it is not contaminated during loading; it is likely to require additional chlorination to ensure it remains wholesome. On-board chlorination plants must be adequately maintained. Ultra-violet treatment systems should not be relied upon as the sole means of disinfection. Reverse osmosis and evaporator units should not be used unless the vessel is sufficiently far from land to ensure that the inlet water is not subject to pollution. Detailed recommendations for freshwater storage and distribution on ships is given in the Department of Transport Notice M.1214 (DT 1986).

5.3.4 Tankers and bowsers

All tankers and bowsers used for the delivery of potable water must be dedicated for that purpose and suitably identified. They should be kept locked and drained when not being used and disinfected and flushed with potable quality water before use. Care must be taken to avoid contamination during use and the quality of the contained water cannot be assured for longer than 24 hours. Water from tankers and bowsers should be boiled before being used for drinking purposes.

5.3.5 Water in containers

The Directive also applies to potable waters sold in bottles or other sealed containers. The quality of water from private supplies should be monitored prior to bottling under the provisions of the Private Supplies Regulations. The quality of water from public supplies is controlled by the Regulations. MAFF has issued regulations which came into force on 1 April 1994 and which set quality standards for water in containers after the point of supply (SI 1994/743, see Appendix E). The MAFF regulations will not control natural mineral waters which are already covered by separate legislation, or waters which are, or have been, issued with a product licence under the Medicines Act 1968.

5.4 Drinks Vending Machines

5.4.1 Machine types

Drinks vending machines may be hand filled or permanently connected to a water supply system.

5.4.2 Sampling

5.4.2.1 Hand filled machines

These machines will be filled from a convenient cold water supply tap using a suitable vessel (for example, a jug). When sampling, take samples from the cup station and the water storage tank.

5.4.2.2 'Plumbed-in' machines

Most plumbed-in drinks vending machines do not have the facility for dumping water to waste and when these machines are used at the start of each day they are likely to draw water which has been standing in the supply pipework overnight. **It is therefore inappropriate** to sample the water entering the machine after running water to waste.

The sampling procedure for plumbed-in machines should be based upon the following sequence;

(i) take a sample from the supply pipe entering the machine **without** running any water to waste; and

(ii) take a vended sample from the cup station of the machine.

For further details and additional guidance see the recommended methods for sampling water supplied to and from machines in Drinks Vending Machines—Code of Practice on Hygiene and Water Supply (AVAB 1987).

5.4.3 Interpretation of results

Standards recommended to its members by the Automatic Vending Association of Britain (AVAB 1989) include the following;

(i) *E. coli* and *Pseudomonas aeruginosa* shall not be present in any 100 mL sample of water dispensed by the machine;

(ii) presumptive coliform organisms should not be present; if they are found in the water supply to the machine this should be investigated further;

(iii) the total viable (colony) counts shall be less than 10000 colony forming units per mL at 22°C and 1000 colony forming units per mL at 37°C in water dispensed by the vending machine, and normally not more than ten times greater than the number in the supply water entering the machine (not flushed to waste).

Chapter 6: Methods and Procedures

6.1 Safety Within the Laboratory

The Health and Safety at Work etc Act 1974 and regulations relating to the Control of Substances Hazardous to Health (COSHH) place a clear duty on management with regard to safety in laboratories to ensure, so far as is reasonably practicable, the health, safety and welfare at work of all employees. Similarly, employers are enjoined not to expose the health and safety of the public to risk.

Implementation of this Act is not solely a management responsibility, for safety continues to rest with heads of departments, although they may delegate certain duties. Equally, individual employees have a duty to co-operate with their employers and to take reasonable care for the health and safety, not only of themselves but also of their fellow workers and of the general public. In order to ensure that the requirements of the Health and Safety Act are being met, the Health and Safety Executive may inspect the premises. The Health and Safety Inspectorate also has the power to close laboratories and, in extreme instances, to prosecute.

Sound technique is the basis of safe microbiological procedures. It is important that all individuals concerned with this work should receive adequate training, and that this training is recorded. It is also important that the necessary laboratory equipment and facilities should conform to accepted codes of safety and good practice (Collins, Hartley and Pilsworth 1974, NWC 1983, HSAC 1991a, 1991b, PHLS 1993).

It is accepted that there is no control over the organisms or agents that may be in water samples received in the laboratory. A distinction is therefore made between the occasional isolation and the deliberate introduction of certain pathogenic organisms to the laboratory.

6.1.1 Classification of micro-organisms, based on risk of infection

Micro-organisms have been categorised into four groups (1–4) in increasing order of hazard to laboratory workers or the community (ACDP 1990). The great majority of organisms handled in the water microbiology laboratory will fall into Hazard Groups 1 or 2. On rare occasions an organism may be isolated that falls into Group 3 (for example *Salmonella typhi*). In such an event all the cultures should be removed to a Containment Level 3 laboratory and any further work continued there.

6.1.2 Basic laboratory (Containment Levels 1 and 2)

Laboratories of this type are intended for work with organisms that fall into Hazard Groups 1 and 2. Ceilings, walls and floors should be non-absorbent, smooth, easy to clean and disinfect, and resistant to the chemicals that will be used in the area. Floors should be non-slip and lighting and heating adequate. A hand basin, other than a laboratory sink, is essential. Benching should be set at a height that is comfortable to work at when seated. Benching should be smooth, resistant to the chemicals used and easy to clean and disinfect.

6.1.3 Containment Level 3 laboratory

Laboratories of this type are intended for work with organisms that fall into Hazard Group 3. They should contain all the features outlined for the basic laboratory but in addition have the following features.

The Level 3 laboratory should be separate from other laboratories, with no access (for example false ceilings, pipe ducts) other than through the entrance door, which should be lockable, and an air transfer grille which is normally fitted to the door. A Type 1 microbiological safety cabinet should be sited in the laboratory away from direct draught of doors and windows which could interfere with its proper function.

A lower pressure in relation to the other laboratories should be maintained in the containment laboratory so that air movement is only one way. All air from this room should be released to atmosphere with no re-circulation to other areas within the laboratory building. During working hours this can be achieved by extraction of air from the room to atmosphere through the safety cabinet with replacement air entering through the door grille.

An international biohazard warning sign should be attached to the door and access should be strictly regulated.

6.1.4 Good practice within the laboratory

Micro-organisms can be ingested by mouth pipetting, from articles that may be contaminated if taken into the laboratory, for example pencils, food, cigarettes and cosmetics, and contaminated hands. Micro-organisms can be taken into the lung by inhalation of bacterial aerosols produced during many routine laboratory operations, for example pipetting, centrifuging, blending. Entry through the skin may result from accidental stabbing with pipettes, contaminated broken glassware or used hypodermic needles and through contamination of cuts and abrasions. Eyes splashed with contaminated material can also lead to infection.

It is therefore important that mouth pipetting is prohibited under all circumstances and that pipetting devices are provided in the laboratory.

Food and drink must never be taken into, stored or consumed in the laboratory. Pipettes, pens, labels etc must never be put in the mouth.

Use of hypodermic needles and glass Pasteur pipettes should be avoided wherever possible and chipped or cracked glass vessels replaced. Glass homogenisers should be wrapped in a cloth and gloves worn.

All equipment should be regularly inspected and any faults and remedial actions noted.

Centrifuge tubes should not be filled to more than two-thirds of their capacity. Work surfaces should be disinfected regularly (usually at the end of the working day) and always after a spill of infectious, or potentially infectious, material. A suitable disinfectant, ready for use should always be available at each work station together with a discard jar for re-usable pipettes and other small items, which should be fully immersed. Discard jars should be emptied frequently and the disinfectant replaced.

Spilt infectious or potentially infectious material and broken culture vessels should be covered with absorbent disposable cloths and soaked in disinfectant, left at least 30 minutes and then cleared into a suitable container.

Properly supported discard bags or bins should be available at each work station, including one that is appropriately robust to contain 'sharps'. No discarded infectious or potentially infectious material should leave the laboratory until it has been sterilized or placed into properly labelled bags for incineration.

All materials that fall into Hazard Group 3 should be processed in a microbiological class 1 safety cabinet sited in a Containment Level 3 laboratory.

Scratches, cuts and abrasions on exposed parts of the body should be covered with waterproof dressings and hands should always be washed after handling infectious material and when leaving the laboratory, after removal of laboratory coats.

Proper protective clothing should be provided and worn correctly fastened at all times when in the laboratory. It should be removed when leaving the laboratory and stored apart from other clothing. It must never be worn in any other areas such as rest rooms and canteens. Separate protective clothing must be provided for use in the containment laboratory.

Further details on containment levels, use of microbiological cabinets and categorisation of pathogens according to hazard are given by the Advisory Committee on Dangerous Pathogens (ACDP 1990).

6.2 Quality Assurance

It is essential that a laboratory is able to demonstrate that all results emanating from it are fit for the purpose for which they are to be used. This is achieved by implementing an appropriate programme of quality assurance. The Regulations define fitness for compliance purposes for the monitoring of drinking water supplies as "capable of establishing, within acceptable limits of deviation and detection, whether the sample contains concentrations or values which contravene the prescribed concentrations or values". Depending on the particular test being used, it is necessary to be able to demonstrate the presence (or absence) of a particular class of micro-organisms in a given sample volume, and to estimate numbers. The detection of small numbers of organisms is particularly important. Because of the ubiquitous nature of most microbes, it is essential to ensure that any organisms that are detected have originated from the original sample and have not been introduced inadvertently during sampling or analysis.

Any effective quality assurance programme must therefore cover the whole process from sample collection to the interpretation and reporting of results. It will also include a system of internal quality control, and participation in a proficiency testing scheme, also referred to as external quality control or assessment, appropriate to water microbiology.

Internal quality control should include procedures for and records of the continual monitoring of working practices, equipment and reagents. These procedures should include the use of blanks, semi-quantitative positive controls, negative controls, reference materials, replicate analyses of samples and control charts.

6.2.1 Definitions

ISO 8402–1986, Quality—Vocabulary, contains the following definitions:

Quality assurance: All those planned and systematic actions necessary to provide adequate confidence that a product or service will satisfy given requirements for quality.

Quality control: The operational techniques and activities that are used to fulfil requirements for quality.

The International Harmonised Protocol for the Proficiency Testing of (Chemical) Analytical Laboratories (IUPAC/ISO/AOAC) contains the following definitions:

Proficiency testing scheme: The system of objectively checking laboratory results by an external agency. It includes comparison of a laboratory's results at intervals with those of other laboratories, the main object being the establishment of trueness. Proficiency testing is designed to assess accuracy.

Internal quality control: The set of procedures undertaken by the laboratory staff for continuous monitoring of operations and results in order to decide whether the results are reliable enough to be released; internal quality control primarily monitors that batchwise trueness of results on quality control materials, and precision on replicate analysis of test materials.

Quality assurance programme/system: The sum total of a laboratory's activities aimed at achieving the required standard of analysis. While internal quality control and proficiency testing are very important components a quality assurance programme must also include staff training, administrative procedures, management structure etc. Accreditation bodies judge laboratories on the basis of their quality assurance programme.

6.2.2 General requirements

Laboratory facilities should be constructed from materials that are robust and easily cleaned. There should be adequate space for dealing with anticipated workloads, with separate areas for the reception of samples, examination of samples, preparation of media and reagents, sterilization and wash-up, and storage (including refrigeration). The laboratory layout should provide a smooth workflow. Benching should provide an adequate and safe linear space for each worker, of which not more than half should be occupied by equipment. Facilities for the examination of samples under special containment may be required when pathogens are being handled. Adequate ventilation and lighting must be provided throughout the laboratory areas.

Microbiological tests must be carried out by suitably trained and qualified personnel in accordance with approved methods. All methods of analysis, including instructions for the preparation and sterilization of reagents and media, must be documented and up to date copies readily available to the analyst. All subsamples, results of culture and other tests, and test reports must be clearly linked with the original sample at all stages of analysis. Staff must be instructed in the safe use of equipment and handling of micro-organisms and the safe disposal of cultures. A laboratory safety manual should be available for easy reference and there should also be a clearly defined system for reporting accidents. The staff should be so organised that, at any one time, each member has a particular area of responsibility and is under the supervision of a single person who is responsible for the overall management of the laboratory. Due regard should also be given to security and confidentiality.

6.2.3 Management

Every laboratory should have an organisational chart showing every member of staff, their roles in day to day performance of tasks and, importantly, the chain of accountability and reporting. The post responsible for the Quality Assurance programme should be clearly defined and each member of staff should have a well-defined job description outlining responsibilities and their limits. A record of staff training should be maintained and regularly updated. There should also be a defined plan for provision of cover for work when staff members are absent.

The reporting system should enable results that require immediate remedial action to be communicated without delay to appropriate persons. Records of laboratory results should be kept for as long as necessary to comply with requirements for archive and audit trails, and should include the date, place, and time of sampling.

The Quality Assurance programme can only be successful if all constituent parts are documented. However, it will provide a valuable source document covering all activities of the laboratory and should only require periodic reviews thereafter.

6.2.4 Laboratory equipment

All microbiological laboratories will require certain basic items of equipment including incubators, water baths, centrifuges, refrigerators and freezers, microscopes, autoclaves, balances, thermometers, pH meters and water purifying apparatus. Such equipment will need regular inspection, cleaning and maintenance and, where appropriate, checking for accuracy.

6.2.4.1 Thermometers

The accuracy of thermometers or other temperature-measuring instruments will require periodic checking against certified reference thermometers. Thermometers should be capable of recording differences of 0.1°C. Ideally, waterbaths and incubators should be provided with continuously recording thermometric instruments. Electronic temperature recording devices are now available that can be used continuously to monitor temperatures inside equipment, and data stored can be reviewed and inspected.

6.2.4.2 pH meters

pH meters will require regular standardisation and should be calibrated before use, using standard buffer solutions. Buffer solutions with pH values in appropriate ranges are available commercially.

6.2.4.3 *Balances*

Balances should be provided with a range of sensitivity that is appropriate for the amount of substance to be weighed. Balances should be kept clean, serviced periodically and their calibration checked and be traceable to national standards.

6.2.4.4 *Pipettes*

Pipettes may be glass (re-usable) or plastic (disposable); all re-usable and a representative sample from each batch of disposable pipettes should be calibrated and be traceable to national standards. Calibrated automatic pipettes may also be used. See also Appendix A.

6.2.4.5 *Water purifying equipment*

A de-ioniser will provide water of satisfactory chemical quality for microbiological purposes. The life of the unit may be prolonged by using distilled water or water that has been produced by reverse osmosis. Containers and pipework should conform to BS 6920 (BSI 1988). The water quality will require regular monitoring with a conductivity meter and to ensure absence of contaminating bacteria the water should be passed through a suitable filter before use. Filters must be renewed at regular intervals.

6.2.4.6 *Sterilizing equipment*

(i) Autoclaves

Autoclaves may be used to sterilize items of laboratory equipment, reagents, media etc and also to render infectious materials safe for disposal. It is important to ensure that the appropriate temperatures and times are attained in all parts of the load. Autoclaves should be checked periodically to ensure that the required time/temperature relationships are attained and each operating cycle should have a chart record on which the cycles are clearly identified. They must be used in accordance with the appropriate safety requirements, particularly when the load consists of fluids in sealed containers. A British Standard, BS 2646 Parts 1, 3 and 5 (BSI 1993), gives guidance on the use and performance requirements for laboratory autoclaves.

(ii) Hot air ovens

These also require regular monitoring of the time/temperature relationships for each operating cycle. They should be fitted with a fan to ensure an even distribution of heat to all parts of the load, see BS 3421 (BSI 1961).

Other forms of sterilizers should be validated before use.

6.2.4.7 *Refrigerators and cold rooms*

Refrigerators should have their operating temperatures (normally between 1 and 5°C) checked regularly and be de-frosted at regular intervals. Cold rooms should be equipped with a continuous recording device. Stored items should be inspected periodically and discarded at the end of their storage life.

6.2.4.8 *Freezers*

These also require regular temperature checks to ensure that they comply with the manufacturers' specifications and should be de-frosted at appropriate intervals. They could be fitted with over-temperature alarms. Their contents will need to be periodically inspected and unwanted items discarded.

6.2.4.9 *Incubators and water baths*

The temperature control of incubators, thermal blocks and water baths is of particular importance in water microbiology. Frequent checks should be carried out to ensure that the temperature does not vary by more than ± 0.5°C when in use. When testing coliforms for thermotolerance at 44°C the temperature should not vary by more than ± 0.25°C. Fill water baths with deionised water and check the water level regularly; clean as necessary. Different loading patterns of incubators will give rise to different temperature attainment patterns in parts of the incubator. This is particularly important in 44°C incubators and standard loading patterns should be documented, validated and adhered to.

6.2.4.10 *Membrane filter holders*

These require checking for leaks and sterilizing before use and cleaning after each use.

6.2.4.11 *Microscopes*

The optics and stage should be cleaned with lens tissue paper after each use. UV microscopes require radiation intensity to be checked, and all microscopes should be serviced, at regular intervals.

6.2.4.12 *Glassware, reagents etc*

All glassware items such as pipettes, flasks, Petri dishes etc used in the preparation of media or the handling of samples will need to be of suitable quality, free from inhibitory substances and adequately cleaned and sterilized before use; in many instances pre-sterilized plastic alternatives can be used. The accuracy of volumetric equipment should be traceable to national standards. Water used in the preparation of media and reagents has to be of high quality, for example de-mineralised or distilled. Chemicals and other substances used in the preparation of reagents and media will also need to be of appropriate analytical quality.

6.2.5 Internal quality control

To ensure that a laboratory is capable of isolating, accurately identifying and enumerating micro-organisms present in a sample, and to avoid contamination during this process with extraneous micro-organisms, it is necessary to have a system of internal quality control. This constitutes submitting to the usual isolation, enumeration and identification procedures samples known to contain micro-organisms similar to those normally sought as well as samples that are sterile. Control tests need to be carried out on all the isolation, enumeration and identification procedures and includes all confirmatory tests. If the procedures are functioning satisfactorily, such micro-organisms will be detected, or in the case of the sterile sample no micro-organisms will be found. The control procedures should be done in parallel with each batch of samples examined.

Control organisms should be freeze-dried wild-type reference strains, obtainable from the National Collection of Type Cultures (NCTC), and reconstituted and diluted with quarter strength Ringer's solution to contain a suitable number. The use of natural water known to contain relevant organisms may also be suitable. All confirmatory tests should include positive and negative controls.

Prepared media should be checked for sterility, and identified by batch reference number. A sample from each batch should be tested to ensure that it will support the growth of appropriate organisms. If selective, it should be checked for the inhibition of growth of unwanted organisms. All media will need to be stored under suitable conditions that will delay deterioration (usually in the dark, and refrigerated if necessary) and discarded at the expiry of their shelf life.

Where membrane filters are used, they should be of suitable quality and tested in the chosen media to ensure that they will support the growth of organisms being sought.

If automatic counting instruments are used these should be tested and calibrated against known values.

6.2.6 Control charts

Quality control charts are used extensively in water chemistry for control of laboratory procedures. It is possible to extend the use of quality control charts to microbiology. However, the natural random variation in the number of organisms found between sub-samples effectively means that wide control limits are necessary. In addition, there may be problems in distinguishing between the chance discrepancies in the number of organisms present in sub-samples and in determining these numbers, ie the true laboratory error. Methodology is being continually developed, for example on the use of split samples to check that processing of a counting method, such as, for example membrane filtration, is being performed consistently. Additional samples, examined in duplicate, may be included with each batch of routine samples to provide within-batch quality control. The duplicate samples can be considered as two halves of a single sample, and the results plotted on a graph on which allowable limits of variation are included. Often, action limits are set at certain values to trigger investigation of the method.

Because of the random nature of the distribution of micro-organisms in water, examination of two halves of the same sample can result in quite wide variation in the counts obtained. For example, if the count reported for the first half of a (duplicate) sample is 5, then the 95% confidence interval (CI) for the count for the second half will be 0–14. The CI for the count in the second half-sample for those counts observed in the first half-sample ranging between 0 to 100 are given in Table 13 in Appendix C.

Water microbiology laboratories should carry out several duplicate analyses regularly and plot the results on a control chart. The two parts of each sample should be treated as separate samples and put through the system in a routine manner. The samples should be positioned at random in the incubator, and these positions changed frequently when different batches of samples are examined. The results should, if possible, be read blind, (ie, the count from the second half-sample should be made in ignorance of the result from the first half-sample), and should be read by the same person.

The first count should be recorded on a control sheet, and the corresponding CI for the second (paired) count entered from Table 13. The second half-sample count is then recorded alongside these figures. If this count falls outside the CI, the fact should be highlighted. The results can also be plotted to give a sequential, visual check on a 'control chart'. If, over a period of time, the second count falls outside the CI on more than 5% of occasions, investigations should be carried out to determine the cause. (Lightfoot et al 1994).

6.2.7 External quality control

Laboratories should participate in a system of external quality control which involves the examination of samples distributed by an external agency. Such samples should be examined as part of the usual laboratory routine. The laboratory's results can then be compared with those obtained by other participating laboratories and so provide an independent check on the quality of the laboratory's performance over time.

6.2.8 Laboratory accreditation

Various schemes are in operation whereby laboratories may be accredited to an approved standard. In addition, the performance of laboratories carrying out analysis for compliance purposes under the Regulations in England and Wales is now subject to scrutiny by the Drinking Water Inspectorate. An agreement has been reached between DWI and the NAMAS Executive on all aspects of drinking water compliance sampling and analysis. Requirements are defined in the Drinking Water Testing Specification. Accreditation of a laboratory to this specification will normally be regarded by the Inspectorate as demonstrating compliance with the relevant regulations.

6.3 Statistical Considerations

A bacterial count from a sample yields valuable information, but it is important to assess the accuracy of this information. There are two main aspects to be considered. First, what can be inferred from this sample about the level of bacterial contamination in the water source from which the sample was taken? Secondly, how accurate was the count for this sample using the method chosen by the laboratory? These two sources of 'sample error' are often confused and this can lead to mis-understandings in the interpretation of a sample count. Generally, it is the first source of sample error which is the more important (Tillett 1993). This potential inaccuracy should be understood both by microbiologists and by those submitting samples for analysis since it affects the interpretation of results and any subsequent actions. The second source of sample error should be brought to the attention of laboratory staff who control the analytical procedures used.

The expression 'bacterial count' will be used to imply the count of organisms being enumerated, for example the total coliform count. The same principles apply to the counting of viruses or other organisms.

6.3.1 Estimating bacterial counts within the water source

The examination of a single sample gives an indication of the bacterial count at a particular point in the catchment or supply and at a particular time. The place at which a sample is collected will have been carefully chosen (see section 6.4) and thus a sample should be typical of the sampling area. The volume of water sampled may not however possess identical characteristics to adjacent volumes of water with respect to microbiological quality. Indeed, only a very small volume of water is examined in the laboratory compared with the body of water in question.

The CI for a bacterial density in a body of water cannot be estimated from a single sample. A mathematical description of the distribution of bacteria is needed before ranges, such as 95% confidence intervals, can be estimated. The only situation where a single sample will give such an estimate is when the bacteria are distributed at random. In this case the appropriate mathematical description is the Poisson formula which has a single parameter (μ = mean = variance) which can be estimated, albeit very approximately, from the single sample count. However, there is no evidence that relevant bacteria are ever randomly distributed in any part of a water system.

For untreated waters, there tends to be enormous variation in microbiological quality (Tillett 1993). In treated waters contamination may be intermittent and organisms may be present in aggregates, often on particulate matter, rather than evenly dispersed. Samples from the same sampling point closely adjacent in time can show large differences in bacterial count.

The parameters describing the distribution of bacteria may change over time and therefore a series of single samples collected at different times cannot be used to estimate confidence intervals for the bacterial contents of the body of water at any one time. They should be used instead to indicate trends over time.

6.3.1.1 Monitoring for trends

Bacterial counts from a series of routine single samples collected from one sampling point at different times should be plotted graphically and studied for trends. With untreated waters, any sustained rises above the usual seasonal range should be investigated. An explanation should also be sought for any single, grossly abnormal count. There is no satisfactory mathematical definition of 'abnormal'. If a plot of routine results is monitored visually, excessively high counts should be obvious, as should changes in the average count, over and above any seasonal variation. With treated waters, the organisms should be absent and any positive findings of indicator organisms should be investigated as described in sections 4.6.2.2 and 4.6.3. Low levels of contamination or erratic contamination will produce some samples which do not yield indicator organisms. The microbiological examination of the water must be seen as only part of the total surveillance procedures. Again, it is impossible to estimate the likelihood of the presence of organisms in the body of water simply by using results from one routine sample.

The sampling of treated waters should be as frequent as practicable in order to increase the chance of detecting any breakdown in microbiological quality of the water and these frequencies are specified in the Regulations. The required sampling frequencies are based on past experience which has aimed at maintaining generally high quality of water at the point of delivery. It would be impractical to sample at such a frequency and with such speed as to prevent all water-borne microbiological disease, but past practice in the UK has kept the numbers of such outbreaks relatively low (Galbraith et al 1987, Benton et al 1989). Constant re-appraisal in the light of new experience (DoE/DH 1990) should aim to reduce the chance of such outbreaks.

6.3.2 Estimating the accuracy of counting bacteria from a single sample

Inaccuracies can arise from inadequate sampling procedures and from prolonged delays or unsuitable storage conditions between collecting the sample and examining it in the laboratory. These will not be considered here, neither will accidental contamination.

Two sources of possible inaccuracy which will be considered are the effects of dilution and the method of counting organisms, both of which come into the category of the second source of error (inherent in laboratory methods) discussed in the first paragraph of section 6.3.

6.3.2.1 Potential imprecision due to dilution of the sample

Samples of treated water should not need to be diluted before examination. Samples of heavily contaminated untreated water may have to be diluted so that a count can be made. If the multiple tube method is used then some tubes, but not all, should show growth. If the membrane filtration method is used then the membrane or plate should not be overgrown.

It is standard procedure to report bacterial counts of indicator organisms as the number of organisms per 100 mL. With undiluted waters, approximately 100 mL of sample are examined (105 mL in the case of the 11-tube series). If the sample has to be diluted (prior to any additional dilution inherent in the multiple tube method) and this dilution is, say for example, 10-fold then only approximately 10 mL of the original sample will be examined. The count obtained is then multiplied by the dilution factor and the new count per 100 mL is now an estimate of the numbers contained in the 100 mL sample. The original 100 mL will have been thoroughly agitated to try to achieve random distribution of the organisms. Therefore the count from the diluted portion can be used to estimate the likely count for the original 100 mL and confidence intervals can be calculated using random distribution theory. The methodology for doing this has been published (Tillett and Farrington 1991) and some examples are shown in Table 2.

The imprecision introduced by dilution is likely to be relatively small compared with the variability in bacterial density in a water source where counts are high enough to require dilution of the sample before examination. It is suggested that the confidence intervals given in these examples in Table 2 are not stated when the results are reported. Quoting such ranges may be mis-understood and taken as a statement about the likely bacterial density in the water source. Table 2 is meant to illustrate the additional imprecision introduced when volumes smaller than 100 mL are actually examined.

Table 2 Estimated count per 100 mL (EC) and 95% confidence intervals (CI) for number of organisms in a 100 mL sample where only a subsample is examined, following dilution.

Organisms observed in the subsample:	10-fold dilution		100-fold dilution	
	EC	CI	EC	CI
10	100	50– 180	1000	480– 1830
50	500	380– 650	5000	3750– 6640
100	1000	820–1200	10000	8190–12200

6.3.2.2 Other imprecision of laboratory methods

Samples of treated waters should contain zero indicator organisms and the presence of very small numbers of such organisms in a sample can be detected and measured with good precision by the methods recommended in this Report. Published work has shown (Tillett 1986) that the multiple tube method is more sensitive than the membrane filtration technique. However, untreated waters may yield moderate or high bacterial counts and in these situations the accuracy with which the count is made by the chosen laboratory method should be considered.

6.3.2.3 *Multiple tube method*

In this method a series of subsamples is taken from the original sample, and processed to ascertain which subsamples show the presence of one or more relevant organisms. A mathematical formula based on laws of probability can be used to estimate the most probable number (MPN) of organisms present, given the number of tubes in the series which show growth (McCrady 1915, 1918, Cochran 1950). These various mathematical approaches have been reviewed by Eisenhart and Wilson (1943), and Cochran (1950) has described the principles involved in the estimation of bacterial densities by dilution methods. Tables have been developed by Swaroop (1938, 1951) which give a greater number of combinations of positive and negative results, some of which in practice should occur only very rarely (Woodward 1957, Man 1975).

A confidence interval was often published with the MPN which demonstrated the uncertainty of the bacterial count for that water sample, as estimated from multiple tubes, not for the water source, although it has often been confused with the latter. For this reason the use of a confidence interval is not recommended.

In the past, the relevant mathematical equations had to be solved approximately. However, the power of modern computers allows exact solution of the probability of counts associated with each dilution-series result (Tillett and Coleman 1985, Tillett 1987). Whilst the re-calculation of most probable numbers shows little discrepancy with previously published values, these new calculations have thrown light on two issues. First, the imprecision of previously published confidence intervals. Second, for moderate or high bacterial density, the multiple tube method does not give a clear 'most probable number'. There is a 'most probable range' (MPR) of counts, all of which are almost equally likely to be as correct as the MPN. All calculations are based on the assumption that the organisms present in the water are evenly distributed and the importance of thorough mixing of the sample cannot be stressed too strongly. Although the multiple tube method is very sensitive for the detection of a small number of indicator organisms, the MPN is not a precise value. Apparent differences between results must therefore be interpreted with caution. It should be appreciated also that variations in bacterial numbers in the water source may be very much greater than any imprecision introduced by the multiple tube method.

Appendix C gives tables of the MPN and the appropriate MPR for a 6-tube series (1 × 50 mL; 5 × 10 mL), an 11-tube series (1 × 50 mL; 5 × 10 mL; 5 × 1 mL) and for a 15-tube series (5 × 10 mL; 5 × 1 mL; 5 × 0.1 mL). Results should be reported with care and it may be appropriate to quote the MPR rather than the MPN in order to illustrate the imprecision of the method. However, it should be clearly stated that the range applies to the sample and not to the water source. When plotting results for trends it will be more practical to use the MPN.

6.3.2.4 *Membrane filtration method*

Counts on membranes are subject to statistical variation and replicate tests on the same sample are unlikely to give the same number of colonies. If 100 mL of sample is filtered and incubated, and then every relevant colony on the membrane counted, and every colony confirmed, then the presumptive and confirmed counts are as precise as the method allows. No statistical imprecision need be considered. If the sample is diluted prior to filtration then the count becomes an estimate of the density in the undiluted sample, as already described. If the presumptive count is taken from only a segment of the membrane then imprecision is introduced comparable to that introduced by diluting the sample, assuming that the segment is typical of the whole plate.

6.3.3 Confirmation

Confirmatory tests of the presumptive colonies present on a membrane should be carried out. Two situations are common which, when there are multiple colonies present, can lead to different approaches being adopted when deciding how many colonies should be tested for confirmation. If the aim is to estimate the count of the relevant colonies then consideration should be given to the imprecision introduced by confirming only a fraction of the number of colonies present, and the examples given below should be studied. The number of colonies tested (for confirmation) should be chosen to give a sufficient level of accuracy. This may require subculturing all colonies on a membrane with fewer than ten presumptive colonies present.

If the aim is to demonstrate the presence or absence of the organism, then a different approach may be chosen, provided that there are no microbiological contra-indications. The presence of the organism is demonstrated as soon as one positive confirmation is made. A laboratory may therefore choose to examine fewer colonies, initially, than when the aim is to estimate the count. However, if the (few) colonies chosen show negative, then the sample cannot be assumed to be satisfactory since other colonies not chosen for confirmatory testing may, if tested, prove positive. Hence other colonies from the stored membrane should be tested. This sequential testing is acceptable only when refrigerated storage of the membrane is not detrimental to the survival of the relevant organism.

If **all** presumptive colonies are tested to confirm their nature then no further imprecision (other than that due to the test method) is introduced when the presumptive count is converted into a confirmed count. If only some of the colonies are tested by confirmatory methods then further imprecision is introduced into the confirmed count. For example, if a presumptive count is made by counting **all** the typical colonies on a plate (to be referred to as N) then it is common practice to make confirmatory tests on some, but not all, of these colonies, unless N is very small. Let n be the number of colonies tested and let x be the number which are confirmed to be of the relevant organism type. The confirmed colony count is then estimated to be xN/n. For example, if 60 colonies were observed on the plate, and 20 were selected at random for testing, and 15 of these were confirmed, then the estimated confirmed count would be $15 \times 60 / 20 = 45$.

It is assumed that the n colonies are selected at random or by some other procedure which ensures that they are a typical subsample of the N colonies and that all N colonies were equally likely to be from the relevant organism group. The conditional probability that y is the true count, given that x colonies have confirmed, can be calculated from an equation of possible combinations:

$$P(x \mid y) = {}^{y}C_{x} \, . \, {}^{N-y}C_{n-x} \, / \, {}^{N}C_{n}$$

The 95% CI for the confirmed count can be found by studying these probabilities for all possible values of y, using the observed value of x. The CI will exclude end of range high and low values of y such that their cumulative conditional probabilities sum to less than or equal to 0.05. This procedure follows the principle described in full by Tillett and Farrington (1991).

As an example, if 10 colonies are observed (N) and only two are tested (ie n=2) then Table 3 shows the complete range of probabilities.

Table 3 Range of probabilities if 10 colonies are counted and only two are tested for confirmation

Given that y =	the probability that		
	x = 0 is	x = 1 is	x = 2 is
0	1.000}	–	–
1	0.800}	0.200}	–
2	0.622}	0.356}	0.022
3	0.467}	0.467}	0.067}
4	0.333}	0.533}	0.133}
5	0.222}	0.556}	0.222}
6	0.133}	0.533}	0.333}
7	0.067}	0.467}	0.467}
8	0.022	0.356}	0.622}
9	–	0.200}	0.800}
10	–	–	1.000}

The brackets (}) show the range of y values which should be included in the 95% CI for the true count. Note that if one of the two colonies tested confirms (x = 1) then all the possible values of y are within the 95% CI, because all the probabilities exceed 0.05.

In general, if only a small number of the total colonies is tested then the CI can be very wide. They also tend to be wider if a substantial number of colonies turn out not to confirm. Some examples are shown in Table 4.

Table 4 Variation in the 95% CI with variation in proportion of tested colonies confirmed

Colonies observed (presumptive count)	Number tested	Number confirmed	Confirmed count	95% CI
10	2	0	0	0– 7
10	2	1	5	1– 9
10	2	2	10	3–10
14	7	5	10	6–12
50	10	5	25	9–41

With treated waters, where the vast majority will yield no or very few presumptive colonies, it is recommended that all colonies should be tested by confirmatory methods to ensure accuracy. For untreated waters it may be worth considering using the presumptive count rather than introduce the additional sampling error which accompanies confirmation of some and not all of the colonies. It should be noted that the practice of confirming a maximum of 10 colonies can still introduce potential imprecision of several fold, especially if the presumptive count is large and some colonies fail to confirm.

6.3.4 Comparing results with recommended standards

The standards for drinking waters concentrate on the presence or absence of indicator organisms and pathogens. No statutory comparisons need be made with actual counts and therefore the potential problem of how to compare actual (for example membrane filtration) and estimated (for example multiple tube) results need not be addressed with respect to potable waters.

6.3.5 Reporting results

The report should be a clear statement of the findings. A further statement on sample error, to qualify these findings, should NOT be necessary for routine samples. The sampling strategy should have been worked out with the aim of acquiring an adequate level of information.

If it is decided that a report for a special sample warrants some statement on accuracy then a clear distinction should be made between sampling error due to variability at the water source and possible error introduced by the laboratory methods.

Absence of organisms and unmeasurably high counts should be reported as follows:

(i) **No organisms detected.** A water sample in which no relevant organisms are detected should be reported as 'none found in the sample examined'. There is no correspondence with the concept of 'limit of detection' which is used with chemical measurements, and therefore an expression such as '< 1 per unit volume' has no meaning.

(ii) **Overgrowth, or all tubes positive.** This means that the laboratory method has failed to estimate the true count because of insufficient pre-dilution. With the multiple tube method it is customary to report this as '> 180' for the 11 tube series or '> 1800' for the 15 tube series, in the appropriate units. In fact, the count could be very much higher than these lower limits. With membrane filtration and other methods the report should be 'count too high to be estimated at the dilution used'.

6.4 Sampling Procedures

6.4.1 Sample containers

Sample bottles should be made from good quality soda or borosilicate glass, or from a suitable pre-sterilized disposable or autoclavable polymer (plastic), and be free from toxic substances. They should be provided by the laboratory and used exclusively for the purpose of microbiological sampling. Glass bottles should not be permitted for sampling on food manufacturing premises or in recreational pool areas, where alternative sterile plastic bottles should be provided. The size of the bottle depends upon the number and type of tests to be carried out. A capacity of about 300 mL should be sufficient and bottles should be fitted with ground glass stoppers or autoclavable or heat-resistant screw caps or silicone rubber liners which will withstand repeated sterilization at 160°C. As an alternative ground glass stoppers can be used in conjunction with disposable strips to prevent jamming. Single-use pre-sterilized bottles can also be used. Glass bottles must be capable of withstanding the conditions required for sterilization. Any containers showing defects should be discarded.

6.4.2 Container preparation

New re-usable bottles and caps should be washed thoroughly with a phosphate-free, non-toxic detergent followed by thorough rinsing with clean and distilled or deionised water, and drained. Used bottles should be rinsed in distilled water. This procedure can be carried out either manually or in a washing machine.

As the water to be examined is likely to contain chlorine or chloramines, sufficient sodium thiosulphate should be added to sample bottles to neutralize these substances. At pH values normally occurring in a water supply, sodium thiosulphate ($Na_2S_2O_3.5H_2O$) at a concentration of 18 mgL^{-1} should neutralize up to 5 mgL^{-1} of free and combined residual chlorine and has no significant effect on the coliform or *E. coli* contents of unchlorinated water on storage (PHLS 1953). It is therefore recommended that at least this level of thiosulphate should be added to all microbiological sampling bottles before they are sterilized, the amount depending on the size of the bottle. Thus 0.1 mL of a 1.8% m/V solution of sodium thiosulphate should be added *pro rata* for each 100 mL of bottle capacity. The sodium thiosulphate should be prepared freshly and added by means of a dispenser used only for that purpose. Bottles should then be capped and where necessary the cap covered with another suitable additional cover, before sterilization. Periodically, after sterilization, a check can if necessary be carried out to ensure the sodium thiosulphate has not been de-activated.

If bottles with ground glass stoppers are used, a strip of paper or foil, for example approximately 75 × 10 mm should be inserted between the stopper and the neck of the bottle before sterilization. This prevents jamming of the stopper and cracking of the glass on cooling. Before sterilization, bottles should be marked with suitable indicator tape capable of showing that the bottle has been through the sterilization process.

Bottles should be sterilized, for example either by autoclaving at 121°C to 122°C for 20 minutes or by heating to 160°C in a fan-assisted oven and holding at that temperature for 60 minutes. Each batch of bottles sterilized should contain a suitable indicator, for example Browne's tube and thermax strips or thermalogs for autoclaves and Browne's tubes and thermax strips for ovens, to show that it has received appropriate heat treatment. The indicators should be placed at appropriate levels within the load being sterilized. The results obtained for all indicators should be checked and recorded. Bottles should only be released for sampling if the sterilizing process is shown to be satisfactory. Sterilizing equipment should be serviced and properly calibrated in terms of time and temperature performance, (see section 6.2.4.6). Sterile bottles should be dated and should be stored in a location provided only for that purpose. A quality system of stock control should be carried out to ensure sterilization and that expiry records are maintained. Sample bottles should be stored in such a way that they should be used within two months of sterilization. Any bottles beyond that expiry date should be washed and re-sterilized as described above. Pre-sterilized tamper-proof bottles bought as such should be used within the shelf-life specified by the manufacturer.

6.4.3 Container labels

6.4.3.1 Pre-scheduled samples

All sample bottles must be adequately labelled. Self-adhesive labels which are easily removed on washing are suitable for this purpose. Labels may be pre-printed or hand written in indelible ink. The information provided should clearly identify the sample and the purpose for which it was taken. As an example the pre-printed label or bar-code should provide the following information;

(i) a reference number;

(ii) the date and time of sampling;

(iii) the reason for taking the sample;

(iv) the specific sample point of the supply;

(v) the type of water, for example raw, filtered, treated etc; and

(vi) for consumers, the location of the sampling point.

Any additional information or amendments to the above should be recorded at the time of sampling.

Residual chlorine should always be recorded. Additional information may be of value when interpreting the results of microbiological analyses and this may be provided on a separate sheet which identifies the sample reference numbers and locations. Such information might include whether the sample tap was properly disinfected and observations on weather conditions, possible sources of pollution or unusual features observed or encountered during sampling. All relevant data should be recorded and given in sufficient detail to allow the exact location to be readily identified, if a repeat sample is required.

6.4.3.2 Non-routine samples

Non-routine samples include consumer complaints, investigational, operational and any other samples which have not been pre-scheduled. Any comments relevant to the sample should also be recorded.

6.4.4 Sample locations

The prime objective is to obtain a sample which is representative of the water being examined. Care must be taken to avoid accidental contamination during sample collection and its subsequent handling and transport to the laboratory. Warning markers should be available and used for problem samples, for example those that might be contaminated with sewage.

6.4.5 Sampling from raw waters

Raw water samples should be taken at the inlet to the works and should be representative of the incoming water quality. Such samples could be taken from inlet channels or from specially fitted raw water taps, which need not be disinfected before samples are taken. Where samples have to be taken from surface waters they should not be taken too near a bank or too far from the point of abstraction. Areas of stagnation in streams and reservoirs should be avoided. Routine samples should be taken as closely as possible from the same sampling location.

Ideally, raw water and water at each stage of treatment should be piped through dedicated sample lines, which must be made from approved materials (Colbourne 1985, BSI 1988), and should be as short as possible. It must be possible to sample without shutting off pumps or disinfection, and 'dead water' must be flushed to waste on each sampling occasion. In practice it may not always be possible to provide dedicated sampling points for all waters and it is then necessary to select sites which provide samples which are as representative as possible of the water to be examined. In certain circumstances it may be necessary to take dip samples. However, stringent precautions should be taken to avoid contamination. See 6.4.12.

Where samples have to be taken from a river or stream, sites should be selected preferably where marked quality changes are likely to occur or where there are important river users causing major discharges or abstractions. Weirs or small discharges which are only very local in effect should be avoided, unless of course, it is established that the effect is not local. Sites should preferably be chosen where flow data are available. When sampling, the banks of rivers and reservoirs should be avoided wherever possible and use made of convenient bridges or gantries to take representative samples from the body of the water.

If sampling is intended to monitor the effect of discharges it should be carried out with careful consideration to the mixing of the discharge and receiving water and its effect on the river or stream, above as well as below the discharge point.

When sampling from boreholes or wells, the sample should be collected from a previously disinfected tap fitted to the rising main before the water passes into any reservoir or cistern. Particular attention should be paid to the cleanliness of the tap. The depth selected for sampling should be determined from hydrogeological and site knowledge. The influence of casings where biofilms may form should also be considered. If the well is fitted with a hand pump, the pump should be cleaned, the mouth flamed and the pump operated for at least five minutes before taking the sample. If the only method of collection available is by means of a pail or can, the sample should be collected in a sterile weighted bottle.

Where adequate sample taps are not available, or where a more comprehensive study of the water quality is required, samples should be collected at a point where water enters a reservoir or lake, and from sampling points representative of the water body. Generally, this will involve the selection of a number of fixed buoy locations, samples being collected from different depths at each point. Sampling from the banks should be avoided because of the likelihood of unrepresentative samples due to disturbance of sediment.

6.4.6 Sampling treated water

6.4.6.1 Outlets from water treatment works

To enable representative samples to be taken for compliance purposes all works should be fitted with proper metal sampling taps which comply with BS 6920, or stainless steel taps, with diaphragm valve and swan neck. No attachments or inserts should be fitted to these taps. Taps should be provided at each outlet main of the works and all taps within the works should be correctly and prominently labelled with the type of sample and flushing time.

The delivery pipework to the sample tap should be as short as possible and in good condition. There should be adequate pressure to obtain the samples. Ideally the sample point should be above ground and there should be adequate drainage to prevent the accumulation of water during flushing. Sample taps should be clean, free from slimes, grease, cleansing agents etc, any of which may affect the microbiological sample. Particular attention should be paid to disinfecting the taps. Constantly running taps and plastic or mixer taps should not be used at sample sites for microbiological compliance monitoring purposes.

6.4.6.2 Service reservoirs and water towers

Covered service reservoirs represent point sources where bacterial ingress can and does happen to varying degrees. For operational purposes, the water undertaker should consider all units with compartmented or multi-reservoir sites as separate reservoirs for sampling purposes unless they are directly interconnected. However, for compliance purposes, water undertakers should be confident that samples taken are representative of all the water leaving each surface reservoir and should take account of such factors as flow patterns, joint inlet/outlet arrangements, common inlet/outlet arrangements. If two or more structures share a common outlet main and have no other outlets then the Drinking Water Inspectorate advise that a suitable single compliance sampling point on the common outlet main will be acceptable for regulatory purposes (DWI 1994). If the reservoir has been out of service for a prolonged period or because of contamination, consideration should be given to the analysis of other indicator organisms, enteroviruses and parasites. Guidance on reservoir inspection, cleaning and maintenance should be followed (WAA 1988).

All service reservoirs should have metal sampling taps with washers, conforming to BS 6920, fitted to each compartment of the tank at points representative of water entering and leaving the reservoir. Sample taps should have no attachments or inserts, should not be exposed to contamination, and should be protected, as far as possible, from freezing and vandalism. Where it is impossible to provide a tap on site, a tap should be provided on the outlet main at the nearest possible point to the reservoir. The delivery pipework should be short and in good condition and it should be possible to flush the sample tap at full bore for the requisite flushing time. Taps should be labelled with a flushing time. When the reservoir is below ground, monitoring of the water temperature during flushing may be useful in indicating when the water from the reservoir is being withdrawn. As an interim measure, pending provision of a proper sampling point, or in the event of lack of access, a good quality consumer tap at the nearest property fed directly off the outlet main could be used. A register of these alternative sites should be maintained and the properties inspected.

6.4.6.3 Sampling taps at consumers' premises

Care should always be exercised over the choice of taps for sampling. Sampling taps should always be clean, free from all attachments and from slime, grease, cleaning agents and other extraneous matter. The taps should be in good repair, and should supply water from a service pipe directly connected to the main. Uniflow mixer taps and taps from a cistern or storage tank should not be used. For microbiological sampling in particular, a knowledge of tap types and the problems associated with them is essential in order to to be able to choose the correct method of disinfection.

6.4.6.3.1 Fixed addresses Samples taken from fixed points may comprise no more than 50% of regulatory microbiological samples. Fixed points for compliance testing should be selected such that they are equivalent to a domestic situation. Hydrants should not be used as fixed sampling sites. Ensure at all times that adequate flushing and disinfection is carried out.

6.4.6.3.2 Random addresses Samples taken from randomly selected addresses must comprise at least 50% of the regulatory samples in any zone in one calendar year. Samples must be taken from taps fed by the rising main. Where the selected premises are not accessible, neighbouring premises should be chosen and labels and log sheets suitably amended. The use of random sampling points may require the sampling of taps which would be considered unsuitable under normal circumstances. Such taps may contain external fittings and may be made of materials which render them impossible to disinfect. Mixer taps where the hot and cold water mix together in the spout should not be sampled. Dual flow "mixer" taps may be sampled. Taps which leak between the spindle and the gland when the tap is turned on should not be sampled. A random sample should not be taken from any tap where the water has been fed through an in-line device such as a water filter or softener. If an unsuitable tap has to be sampled, details of any cleaning or disinfection procedure, together with any comments should be recorded on the label or log sheet.

It should be emphasised to staff taking samples that the exact randomness of a sample is less important than the suitability of the tap from where the sample is taken. It is more appropriate to sample a property (containing a suitable tap) adjacent to or in the proximity of an address which is pre-scheduled and chosen at random but which contains a tap considered unsuitable.

6.4.7 Order of taking samples

When sampling from fixed sites, such as raw water taps, treatment works taps, taps on service reservoirs or fixed points in distribution, or when sampling from a random site for regulatory purposes or in response to a consumer enquiry, the order of sampling should be as follows:

1 Samples that are required to be taken before flushing.

2 Flush for a specified length of time.

3 On-site chemical and physical tests, for example chlorine residual, temperature, pH.

4 Physical and chemical samples required to be taken after flushing, and biological samples, including invertebrate samples if required.

5 Disinfect the tap (section 6.4.8).

6 Bacteriological sample.

Swab samples, if required, can be taken first or last according to the reason for sampling (see section 6.4.8.4).

6.4.8 Sampling procedure—sampling from taps

All treated water sample taps should be disinfected before being sampled for microbiological analysis, except when investigating a consumer complaint (see section 4.6.2.1), when a pre-disinfection sample may also be taken, if there is justification.

6.4.8.1 Disinfection by flaming

This procedure can be carried out on all metal taps. Turn the tap on and leave to run at a uniform rate for 2–3 minutes. Turn the tap off and flame thoroughly starting at the nozzle and working back to the body of the tap until the water held in the spout boils. Care should be taken that very hot water, which may spurt out of the tap during flaming, does not cause injury. If the spout drains dry when the tap is turned off, heat the full length of the spout such that the first issue of water boils when the tap is turned back on. A propane or butane burner can be used for flaming as it should produce a tight controllable flame. Methylated spirit flames should not be used as they are not hot enough and are difficult to control. Ensure that flammable or heat-sensitive items, such as curtains, are moved away from the fire zone. After flaming, run the tap to waste until the water is cool.

6.4.8.2 Disinfection by using sodium hypochlorite solution

This procedure can be carried out on plastic taps, mixer taps and metal taps where flaming is not appropriate. Remove all external fittings. Accumulations of grease and slime should be removed from the tap with a clean sterile swab. Turn the tap on and leave to run at a uniform rate for 2–3 minutes. Disinfection can be carried out by swabbing the outside of the tap, and as much of the inside as possible, with a 10% solution of commercial sodium hypochlorite solution (to give 1% available chlorine). Alternatively, use a wash bottle or similar device filled with a 1.0% solution of sodium hypochlorite to spray outside the tap and inject inside the tap spout. Disinfection swabs should fit snugly within the spout to ensure maximum contact with all surfaces. **Sodium hypochlorite is highly corrosive and must be handled with great care. If contact with skin or clothes occurs, wash immediately with copious amounts of water.** Leave for 2–3 minutes to allow the sodium hypochlorite to disinfect the tap. Run the water to waste for a sufficient period to ensure that all the sodium hypochlorite is removed before taking the sample.

6.4.8.3 Taking the sample

Take the sample in a sterile bottle. Hold the bottle in one hand and remove the stopper or cap with the other. The sample bottle should never be rinsed out or the closure put on any surface. Fill the bottle, leaving a small air gap, from a gentle stream and avoid splashing. Do not change or alter the flow rate of the water leaving the tap as the bottle is being filled. To do so may cause debris or other material to become dislodged from the system and enter the sample bottle. Replace the stopper or cap immediately. Care must always be taken not to touch the top of the bottle during removal or replacement of the cap. Secure the cap and shake the sample before returning it to a cool box. If accidental contamination is suspected, the sample must be discarded and a fresh, sterile bottle used to take a further sample.

6.4.8.4 Bacteriological swabs

Additional information about the microbiological state of the tap can sometimes be obtained by the use of a sterile swab. It can be helpful to swab sample random addresses and all customer complaints; the swab samples may be taken before and/or after all other sampling has been completed. After removal from its container, the swab should be rubbed around as much of the inside surface of the tap as possible and then carefully replaced in its container. The container should be clearly identified and labelled with appropriate details.

6.4.9 Sampling from hydrants

Regulatory or compliance samples must not be taken from hydrants or standpipes. Remove the hydrant box lid and remove any debris from the area of the connection. Bail out any accumulated water to well below the bottom of the threads. Unscrew the hydrant cover and open the hydrant valve a small amount, sufficient to flush any accumulated debris from the hydrant. Run to waste approximately 5 litres of water.

Unwrap the standpipe, which should be transported in a clean plastic bag, and connect it to the hydrant. Open the standpipe tap smoothly; abrupt or rapid opening can cause turbulence in the main and stir up sediment. Flush water through the standpipe until the water runs clear. Turn off the water. If net samples for animals or similar samples are required, they should be taken at this point.

Remove the standpipe and bail out 200 mL of water from the hydrant bowl. Add 100 mL of a 1.0% solution of hypochlorite. Care should be taken and any safety requirements observed. See section 6.4.8.2. Replace the standpipe and open the valve slightly, until chlorinated water first discharges. Turn the water off. Wait for 10 minutes before flushing the chlorinated water out of the hydrant. Flush for one minute and, without turning off the water, measure the chlorine content of the water. Continue flushing until the chlorine content is below 0.5 mgL^{-1} or until the level is no higher than the chlorine level in the mains water. Without turning the water off, take a sample of the water, avoiding either the bottle or the top touching any surface. Turn off the water, remove the standpipe and replace the hydrant cap and the box lid.

6.4.10 Sampling from bowsers

All taps and hatches should be checked to ensure no damage has occurred due to vandalism and that the bowser contains water. Check the free and total residual chlorine from the bowser. Run the tap to waste for about 10 seconds. Disinfect the tap as described in 6.4.8, flush and sample.

6.4.11 Sampling new mains, renovated mains and repaired mains

If applicable, follow the procedures in section 6.4.9 (sampling from hydrants). Other details can be found in 'Operational Guidelines for the protection of Drinking Water Supplies' (WAA 1988).

6.4.12 Sampling for investigational purposes—dip sampling

Dip sampling should only be carried out for investigational purposes and not on a routine basis or for regulatory purposes. A dip sample may be taken for microbiological analysis using a sterile dip bottle. Sterile dip sample bottles can be prepared by using a wide mouthed 500 mL sample bottle with a wire attached. The sample bottle can be wrapped in suitable material and autoclaved with the lid wrapped in tin foil. Clean string can be attached to the wire and then the outer wrapping from the bottle removed. The wrapped lid can be held until the bottle has been dipped and filled. The lid is then unwrapped and carefully placed on the bottle. Dip samples may also be taken using a sterile weighted sampling device or a sample bottle attached to a long pole.

6.4.13 Drinks vending machines

See section 5.4.2 for details of how to take samples from drinks vending machines.

6.4.14 Transport and storage of samples

A review of published studies shows that indicator organisms from different water sources can behave differently in transit (Tillett and Benton 1993). Therefore samples should be delivered to the laboratory and analysed as quickly as possible after collection. This particularly applies to bacteriological samples where every effort should be made to start examination within six hours of collection.

Samples for microbiological analysis should be kept cool (2–10°C) during transportation. Consideration should be given to the manufacture of specially designed containers, or the provision of mechanically refrigerated compartments in sampling vans. Containers used for transporting microbiological samples should not be used for other purposes but if this is unavoidable, care must be taken not to contaminate the samples. Containers must be cleaned and disinfected regularly. Where logistics do not allow examination within six hours, storage may be prolonged and samples may be examined up to twenty four hours after collection provided that samples are kept in the dark and cool (2–10°C) (Guidance Document). This procedure should be regarded as a 'last resort' and should only be undertaken in exceptional circumstances. If a sample is examined after a much longer storage period and is found to contain undesirable organisms, for example a faecal coliform, this is a significant finding since the organism must have been present earlier. However, failure to find an organism after an extended storage period does not rule out the possibility that the organism was present at an earlier stage. For this reason it is emphasised that all samples should be examined as soon as practicable after collection. This advice has been confirmed by the findings of a research contract funded by DoE (PECD 7/7/413). This research showed that for raw waters a statistically significant reduction occurred in faecal colony counts after twenty four hours storage compared to the analysis after six hours storage.

Chapter 7: Bacteriological Methods

7.1 Introduction

This chapter describes in detail the various procedures for the microbiological examination of water. The purpose of this detailed description is not to prescribe a rigid set of rules from which any deviation is to be deprecated, but rather to specify techniques which are likely to be accompanied by acceptable experimental error yet give reproducible results. Changes to the methods must be fully documented and evaluated in order to establish whether the performances of the modified methods remain acceptable. The importance of proper quality assurance has already been emphasised. The methods described in this section of the Report have been used widely for many years in the assessment of the wholesomeness of drinking water, and are accepted as appropriate for compliance monitoring in the UK. Changes in technique and the introduction of new methods can only be justified by the demonstration that the changes provide equivalent or better performance in terms of sensitivity and specificity.

Details of the preparation and sterilization of necessary glassware are described in Appendix A and methods for the preparation of media and reagents in Appendix B. These methods should be adhered to if uniform results are to be obtained. Whilst the use of certain standard commercial preparations is occasionally recommended, no endorsement of any particular manufacturer is suggested. The need for regular monitoring of the performance and accuracy of laboratory equipment and for testing the quality of reagents and media is, however, greatly emphasized. Particular attention should be paid to the control of operating temperatures of incubators and water baths.

7.2 Laboratory Hygiene

In water examination, particular significance is attached to the presence of very small numbers of coliform organisms and *E. coli*. As these organisms are very common in humans and their environment, and are cultivated routinely in the laboratory, special precautions are needed to avoid accidental contamination of samples. These have been discussed earlier, and especially in Chapter 6, but the need to conform to good laboratory practice at all times is stressed.

7.3 Methods for the Detection and Enumeration of Indicator Organisms

Two procedures, membrane filtration, and the multiple tube test, also known as the most probable number (MPN) or dilution method, are available. Both these techniques can be used to detect and enumerate indicator organisms, the media and incubation conditions differing according to the organism sought. Each method is therefore described in detail together with its application to the test for coliform organisms and *E. coli*; the variations necessary for the other indicator organisms are described in later sections. The preparation and dilution of samples are the same for every test and these are described first.

7.4 Preparation of Samples

To facilitate mixing, an air space should be present in the sample bottle. Invert the sample bottle rapidly several times. Hold the bottle in one hand and remove the stopper or cap with the other but at the same time retaining it. Flame the mouth of the bottle. Make any dilutions required at this stage.

The sample volume should be sufficient to ensure all routine examinations can be carried out. Any excess sample not required may be stored in a refrigerator until the initial examination has been completed. This sample can then be discarded, or used for further tests in the event of an unexpected high count or other mishap.

7.4.1 Diluent

Use sterile quarter-strength Ringer's solution for all dilutions, see page 119.

7.4.2 Making the dilutions

Measure out a known volume, for example 90 or 9 mL of diluent into sterile dilution bottles or tubes. Alternatively, use volumes of diluent pre-sterilized in screw-capped bottles. In this case it should be recognised that some bottles may lose diluent on storage. Volumes should be checked and any bottles obviously containing incorrect amounts must be discarded.

Make one, or more, ten-fold dilutions by transferring one volume of sample to nine volumes of diluent.

Using a fresh, sterile pipette each time, repeat the process as often as necessary to ensure the correct dilution range has been covered. Carefully mix each prepared dilution before further dilution or before use. Prepare sufficient quantities of each dilution to enable all tests to be carried out.

7.5 Membrane Filtration Method

7.5.1 Principle

In the membrane filtration method a measured volume of the sample is filtered through a membrane composed of cellulose esters or equivalent. The pore size is such that the organisms to be enumerated are retained on or near the surface of the membrane which is then placed, normally face upward, on a differential medium selective for the indicator organisms sought. This may be either an agar medium or an absorbent pad saturated with broth. After a specified incubation period, it is assumed that the indicator organisms retained by the membrane will, depending on the medium used, form colonies of characteristic morphology and colour. Other organisms are either inhibited or can be distinguished by their colonial appearance. The colonies of the organism sought are counted and the result, taking into account any dilutions made, is expressed as the number per 100 mL of the sample. This presumptive count is then confirmed by subculturing all or a representative number of colonies formed.

7.5.2 Filtration apparatus

The membrane filtration apparatus consists of a base supporting a porous disc. The filter funnel, which should be graduated, is secured to the base by means of screw-threads, clamps or a magnet. The filtration apparatus is connected to a vacuum source. For the examination of large numbers of samples, multiple filtration units may be used. The filtration apparatus should be sterilized on a regular basis between batches of analyses. Spare funnels as required can be disinfected by immersion in boiling or nearly boiling distilled water for at least one minute between samples. After disinfection, each funnel should be placed in a stand and allowed to cool before use. Alternatively, a fresh pre-sterilised funnel may be used for each sample.

7.5.3 Membranes

Membrane filters, for example 47 mm in diameter, with a rated nominal pore-size of 0.45 μm or membranes which have equivalent filtration characteristics, are recommended; these will retain almost all the bacteria mentioned in this Report. A pore size of 0.2 μm is, however, recommended for the isolation of campylobacters (section 8.6.6.1). The use of membranes with grid-marks is recommended to facilitate counting. It is necessary to check periodically that the membranes used are suitable for the organisms sought in the waters examined. Quality assurance is important and membranes must be free from toxic substances. When membranes with grid-marks are used, bacterial growth should not be inhibited or stimulated along the lines.

Membranes should be obtained pre-sterilized and should not be re-used. Membranes should not be used beyond their shelf life date.

7.5.4 Absorbent pads

Absorbent pads of at least the same diameter as the membranes and approximately 1 mm thick should be used. The pads should be made of high quality paper fibres, and be uniformly absorbent and free from any toxic substances which could inhibit bacterial growth. Absorbent pads are available pre-sterilized or they can be sterilized, preferably at 121°C for 20 minutes, either in containers or wrapped in waterproof paper or foil.

Place each individual pad with aseptic precautions in a sterile Petri dish.

7.5.5 Media

Details of media recommended are given in Appendix B. The media used with membrane filters differ in composition from those for the multiple tube method because membranes selectively absorb certain substances; there is also a differential absorbent capacity between the nutrient pad and the membrane. Dehydrated media should be prepared according to the manufacturers' instructions. Alternatively, the media may be prepared in the laboratory from approved high quality constituents. Sterile broth is dispensed aseptically on to each filter pad in sterile Petri dishes or similar containers; this may be done conveniently with an automatic pipette. The pad should be saturated with broth. When agar medium is used, the surface of the agar should be completely smooth and dry.

7.5.6 Incubators and water baths

Accurate temperature control of incubators is essential, see section 6.2.4.9. When resuscitation techniques involving pre-incubation at lower temperatures are required, manual transfer between different incubators can be avoided if an incubator with fan-assisted air circulation and fitted with a time control to give automatic switching between the temperatures is used. Incubators of this type function adequately only when the lower temperature is at least 5°C above ambient. When incubators are required to operate in an environment where the ambient temperature may approach within 5°C of the lower temperature, the use of a cooled incubator is essential. Instead of incubators, dual time-temperature controlled water baths with circulation may be used. With water baths, it is essential to use suitable water-tight immersible containers to hold the Petri dishes.

7.5.7 Procedure

7.5.7.1 Preparation of samples

The volumes and dilutions of samples should be chosen so that the number of colonies to be counted on the membrane lies, if possible, between 20 and 80. With some waters, it may be advantageous to filter different volumes so that the number of colonies on one of the membranes is likely to fall within this range. For treated waters, filter 100 mL of the sample; for polluted waters either filter smaller volumes or dilute the sample before filtration. When the volume to be filtered is less than 10 mL, add at least 20 mL of sterile diluent to the funnel before addition of the sample to aid uniform dispersion of the bacteria over the entire surface of the membrane during filtration.

7.5.7.2 Filtration procedure

Place the disinfected filtration apparatus in position and connect to a source of vacuum, with the stopcock turned off.

Remove the funnel and place a sterile membrane, grid-side upwards, on the porous disc of the filter base. Grasp only the edge of the membrane filter with flat-ended sterile forceps. Replace the sterile funnel securely on the filter base. Pour or pipette the required volume of sample into the funnel. Open the stopcock and apply a vacuum of about 65 kPa (500 mm of mercury) and filter the water slowly through the membrane. Close the stopcock as soon as the sample has been filtered so that as little air as possible is drawn through the membrane.

Remove the funnel to a boiling water bath if it is to be re-used and transfer the membrane carefully either to a pad saturated with medium or to a well-dried agar plate. Ensure that no air-bubbles are trapped between the membrane and the medium.

Pour off any excess medium from the saturated pad, either before or after the membrane is placed in position. If this is not done, confluent growth may result.

For different volumes of the same sample, the funnel may be re-used without boiling provided that the smallest volumes of sample are filtered first.

For different samples, take a fresh pre-sterilised funnel or remove a funnel from the boiling water bath, allow to cool and repeat the filtration process.

After filtration of each sample, disinfect the funnel by immersion in boiling distilled water for at least one minute. During the filtration of a series of samples the filter base need not be sterilized unless it is contaminated or a membrane is damaged.

Do not alternate the filtration of known polluted samples with those of treated samples through the same apparatus.

Filter all samples of chlorinated water and those expected to give negative results first, followed by those known or suspected of being polluted. Alternatively, a separate membrane filtration apparatus can be reserved for all chlorinated samples and another for polluted samples.

When funnels are not in use they should be covered with a sterile lid or a sterile Petri dish lid.

The time between the end of the filtration step and the beginning of the incubation stage should be as short as possible and no longer than 2 hours.

7.5.7.3 Incubation and examination

When using pads soaked with liquid medium, place the Petri dishes with the membranes inside a container with a tightly fitted lid to prevent drying out. Alternatively, a polythene bag may be used and carefully folded over, tied or sealed. For anaerobic organisms, special incubation conditions are required.

Incubate the membranes at the appropriate temperature and for the time specified for the organism sought.

After incubation, count the characteristic colonies on the membrane, if necessary with a hand lens. Make a note of the numbers of each morphological type. As colours are liable to change on cooling and standing, count the colonies within a few minutes of removal from the incubator or water bath.

Express the result as the number of presumptive indicator organism per 100 mL of sample (see section 6.3.5).

7.5.8 Advantages and limitations of the membrane filtration method

7.5.8.1 Advantages

The outstanding advantage of the membrane filtration technique is the speed with which results can be obtained as direct counts. For example, presumptive coliform and *E. coli* counts may be available within 18 hours. This enables a quicker response to be made with respect to the remedial action that may be required. Individual colonies can be available for confirmatory testing after 18 hours.

There is considerable saving in technical labour and in the amount of media and glassware required compared to other techniques.

The conditions of incubation can be varied easily to encourage the growth of attenuated or slow-growing organisms.

False-positive reactions that may occur with some media in the multiple-tube test for coliform organisms due to the growth of aerobic and anaerobic spore-bearing organisms, or to mixtures of organisms, are unlikely to occur with membranes.

7.5.8.2 Limitations

In the membrane filtration method, gas production, is not detected. This however is less critical in view of the recognised change in the importance of gas production in relation to the definition of coliforms, see section 3.2, but may be a consideration with respect to other organisms.

Membranes are unsuitable for use with waters of high turbidity. In these circumstances the membrane will become blocked before sufficient water can be filtered, while the accumulated deposit on the membrane may inhibit the growth of indicator organisms.

The membrane method is also unsuitable for water containing only small numbers of the indicator organisms sought in the presence of many other bacteria capable of growth on the media used (see, for example section 7.7.5).

7.6 Multiple Tube Method

7.6.1 Principle

In the multiple tube method measured volumes of sample, or of one or more dilutions, are added to a series of tubes containing a liquid differential medium. It is assumed that on incubation each tube which received one or more test organisms in the inoculum will show growth, and characteristic change produced by the organism sought when growing in the medium used. Provided that negative results occur in some of the tubes, the most probable number of organisms in 100 mL of the sample can be estimated from the number and distribution of tubes showing positive reactions. Confirmation that positive reactions are due to the growth of the specific indicator organism sought can be obtained by subculture to tubes of confirmatory media, some of which may need to be incubated at higher temperatures, depending on the organism sought. Positive reactions in the confirmatory tests yield further information about the types of organism present. The presumptive and confirmed counts are calculated by reference to probability tables (see section 6.3).

The multiple tube method is particularly suitable for the examination of sludges and waters containing sediment.

7.6.2 Procedure

7.6.2.1 Choice of volumes for inoculation

With waters expected to be of good quality, add 1 × 50 mL volume and 5 × 10 mL volumes of the sample to equal volumes of double strength medium.

With waters of doubtful or unknown quality, use 1 × 50 mL volume, 5 × 10 mL and 5 × 1 mL volumes; add the 1 mL volumes to 5 mL of single-strength medium.

With more polluted waters include also 5 × 0.1 mL volumes (5 × 1 mL volumes of a 1 in 10 dilution) and omit the 50 mL volume of the sample.

With heavily polluted waters, additional dilutions of a hundred-fold, a thousand-fold or even higher, may be needed to give some negative reactions necessary for the MPN estimation.

7.6.2.2 Inoculation of the culture medium

With a sterile pipette add 1 mL volume of sample to a tube containing 5 mL of single strength medium, and 10 mL volume of sample to a tube containing 10 mL of double strength medium. Measure 50 mL volume of sample into a screw-capped bottle containing 50 mL of double-strength medium.

7.6.2.3 Incubation and examination of the cultures

Incubate the inoculated tubes at the specified temperature, and examine them at set intervals, according to the indicator organism being sought.

Count the number of tubes of each volume showing a positive reaction and by reference to the appropriate tables in Appendix C, calculate the MPN of organisms present in 100 mL of the sample. For example, in a 15-tube test with 5 × 10 mL, 5 × 1 mL and 5 × 0.1 mL volumes, if the number of tubes showing a positive reaction in each set of five tubes is 3, 2 and 0 respectively, then the MPN is 13 organisms per 100 mL. At this stage carry out any confirmatory tests that may be required.

When using dilutions, choose a consecutive series of three ten-fold dilutions which show some positive and some negative reactions and multiply the MPN by the appropriate dilution factor as described in Appendix C.

7.7 Count of Total Coliforms and *Escherichia coli* by Membrane Filtration

7.7.1 Introduction

Tests for coliform organisms and *E. coli* are the most important routine microbiological examinations carried out for drinking water as they provide the most sensitive method for assessing the effectiveness of disinfection and for detecting faecal contamination.

7.7.2 Principle

Isolation of organisms on a membrane placed on an absorbent pad saturated with broth containing lactose and phenol red as an indicator of acidity, with subsequent confirmation of acid production from lactose and, where necessary, oxidase reaction and indole formation.

7.7.3 Definitions

In the context of the method, organisms which are oxidase negative produce acid from lactose and form all shades and sizes of yellow colonies on membranes after incubation for 4 hours at 30°C followed by 14 hours at 37°C are regarded as total coliform organisms.

These are considered to be members of genera or species within the Family Enterobacteriaceae, capable of growth at 37°C, that possess ß-galactosidase. This definition will include anaerogenic strains. Coliforms are, by definition oxidase negative. The following genera have been isolated in routine practice:

Citrobacter, Enterobacter, Escherichia, Hafnia, Klebsiella, Serratia, Yersinia.

Organisms which produce acid from lactose after incubation for 4 hours at 30°C followed by 14 hours at 44°C are presumptive faecal coliform bacteria, of which the majority of isolates may subsequently confirm as *E. coli.*

The term "faecal coliform" is not precise and is used to denote a coliform of faecal origin; it is used in this sense in the Directive and in the Regulations. The Directive implies (and the Guidance Document states) that faecal coliforms are those capable of growth at 44°C, ie are thermotolerant. In some documents thermotolerant coliforms are referred to as faecal coliforms. However, as will be noted not all thermotolerant coliforms are of faecal origin nor are all faecal coliforms thermotolerant. Thermotolerant coliforms should therefore be regarded as "presumptive faecal coliforms".

For the purposes of water examination *E. coli* has historically been regarded as a member of the Family Enterobacteriaceae which ferments lactose or mannitol at 44°C with the production of acid and usually gas within 24 hours, and which produces indole from tryptophan. Most strains produce ß-glucuronidase. Strains which possess these characteristics at 37°C but do not express them at 44°C and strains which are anaerogenic (do not produce gas) may also be *E. coli.* When identified as *E. coli* they have the same sanitary and operational significance with regard to their faecal origin, and, in order to resolve some confusion in terminology, they are regarded in this Report as faecal coliforms. The Drinking Water Inspectorate advised the committee responsible for revising this manuscript that for UK regulatory purposes, confirmed *E. coli* can be regarded as faecal coliforms. For practical purposes, examination for organisms satisfying the definition of coliforms and producing indole from tryptophan at 44°C can be considered as examination for *E. coli.*

Under certain circumstances, *Aeromonas* spp., may grow under the conditions of the test for total coliforms, but may be recognised because they are oxidase positive. The oxidase test is essential for the confirmation of coliforms. Confirmed total coliforms and *E. coli* are oxidase negative.

7.7.4 Choice of medium for isolation and enumeration

Membrane lauryl sulphate broth (MLSB) (PHLS/SCA 1980a) is recommended for the isolation and enumeration of total coliforms and *E. coli*. This is available commercially as a complete dehydrated medium.

7.7.5 Procedure

For each sample, place an absorbent pad into each of two empty sterile Petri dishes. Add sufficient MLSB to saturate the pad, allow to soak in and pour off any excess fluid.

Prepare any necessary dilutions and filter as described in section 7.5.7. Set up two membranes for each sample and place on the pads soaked in MLSB.

Incubate both membranes at 30°C for 4 hours then transfer one membrane to 37°C for total coliforms and the other to 44°C for *E. coli*. Use incubators or waterbaths for incubation as described in section 7.5.6. Accurate temperature control and even temperature distribution are essential, especially for *E. coli* at 44°C. False positive results will be obtained at temperatures below that recommended and some strains will fail to multiply at higher temperatures. Incubate the membranes at 37°C or 44°C for 14 hours to give a total incubation time of 18 hours. If an early indication of a result is required urgently, the membranes may be examined for presumptive positive results after a total incubation time of 12 hours but must be returned to the incubator for the full period of 18 hours before results can be regarded as negative. From a water treatment point of view it may be convenient to incubate a single membrane at 37°C. In this case an immediate operational response should be made to **any** presumptive positive result and should be treated as *E. coli* until the confirmatory tests for coliforms and *E. coli* have been completed. It should be assumed that any presumptive indication is treated as *E. coli* and remedial action taken.

After incubation, examine the membranes under good light, if necessary with a hand lens. Count all yellow colonies (however faint) irrespective of size within a few minutes of being removed from the incubator. Colours are liable to change on cooling and standing. It is important to note whether pink colonies are present in numbers which may interfere with the growth of coliforms. The detection of acid production is influenced by the pH of the medium, thus it is important that the medium is of the correct pH. If the growth of pink colonies is considered to be such that they may be obscuring lactose-fermenting colonies, a further sample should be taken and examined by membrane filtration; appropriate dilution and presence-absence methods should be considered (section 7.8.8).

7.7.6 Confirmatory techniques

7.7.6.1 Colonies chosen for subculture

With raw waters, it is not essential to confirm presumptive total coliform and *E. coli* results except in special circumstances, for example borehole waters.

For treated waters, subculture a suitable number of yellow colonies (however faint) depending on the purpose intended and accuracy required (see section 6.3.3). If the aim is to estimate the number of organisms present, then for the greatest accuracy, all colonies should be examined if fewer than ten are present, and at least ten colonies if more are present. Colonies should always be chosen at random, but to avoid any bias from, for example, unconscious choice of similar colonies, all the colonies in a randomly chosen segment of appropriate size should be examined. Where a number of colonies is clearly distinguishable a note of the number of each morphological type should also be made. The number of confirmed colonies on the membrane can then be interpreted. If confluent growth is obtained, any sample retained in the refrigerator (section 7.4) may be re-examined using an appropriate dilution to enable isolated colonies to develop.

Whenever colonies are subcultured for confirmation, they should be tested for confirmation both as total coliforms and as *E. coli*, whether initially isolated at 37°C or 44°C. This is important because presumptive coliform colonies at 37°C may confirm as *E. coli*, and presumptive *E. coli* colonies isolated at 44°C may not confirm as *E. coli* but may nevertheless confirm as coliforms. It is particularly important when colonies are present on only one of the two membranes incubated at 37°C and 44°C. It must also be recognised that *E. coli* is itself a coliform, and that if the confirmed *E. coli* count from the 44°C membrane exceeds the confirmed coliform count from the 37°C membrane then the confirmed *E. coli* count must also be recorded as the total coliform count.

If there are colonies present on only one of the two membranes incubated at 37°C and 44°C they should be subcultured for confirmation for both total coliforms and *E. coli*. In the case where, for example, 0 coliforms have been isolated at 37°C but 2 confirmed *E. coli* at 44°C then results should be reported as 2 coliforms and 2 *E. coli*.

7.7.6.2 Confirmation of total coliform organisms

Subculture each colony to be tested from the membrane to Lactose Peptone Water (LPW) and incubate at 37°C. After 6 hours at 37°C the LPW cultures may be subcultured to MacConkey Agar (MA) and Nutrient Agar (NA) plates to check for purity and colonial appearance. Large isolated colonies may be subcultured to MA and NA direct from the membrane. Incubate the LPW cultures at 37°C and examine for acid production after 24 hours and, if the results are negative, after a further 24 hours.

Incubate the MA and NA plates at 37°C for 24 hours and carry out an oxidase test (section 7.8.7.2) on colonies only from the NA plate. Pure cultures are essential for the oxidase test and it may be necessary to make further subcultures. These should be made from characteristic coliform colonies on the MA plate, (section 7.8.7.3). In-situ oxidase tests carried out directly on colonies on the membrane are not recommended. Oxidase may diffuse from oxidase-positive colonies to adjacent oxidase-negative colonies and oxidase production may be inhibited by acid produced from lactose. Also, further subculture may be unsuccessful due to the toxic effects of the oxidase reagent.

Coliforms are oxidase negative, possess β-galaetosidase and produce acid from lactose within 48 hours at 37°C. Further identification may be carried out using characteristic colonies on MA by means of appropriate biochemical and other tests (Cowan 1993). Commercial test kits may be used.

7.7.6.3 Confirmation of E. coli

Subculture yellow colonies from the membranes incubated at 37°C and 44°C to MA and NA, to LPW (containing a Durham tube if information regarding gas production is required or considered relevant), and to Tryptone Water (TW). Incubate all media at 44°C for 24 hours. *E. coli* is oxidase negative, produces acid (and usually gas) in LPW, and indole in TW. These tests are detailed in sections 7.8.7.1 to 7.8.7.3. Tests for ß-glucuronidase may assist in the early confirmation of *E. coli* (Feng and Hartman 1982, Kampfer, Rauhoff and Dott 1991). Again, suitable commercial test kits may be used.

7.8 Count of Total Coliform Organisms and *Escherichia coli* by Multiple Tube

7.8.1 Principle

Isolation of organisms in a liquid medium containing lactose, a suitable pH indicator for the detection of acid production and, if information is required or considered relevant, a Durham tube for the detection of gas production.

7.8.2 Definitions

See also sections 3.2. and 7.7.3. In the context of the method, total coliforms produce acid from lactose within 48 hours at 37°C.

E. coli is a member of the Family Enterobacteriaceae which produces acid, and usually gas, from lactose or mannitol at 44°C and which produces indole from tryptophan. Some strains are anaerogenic (non gas producing).

Coliforms can be identified to presumptive level without an oxidase test but demonstration of a negative oxidase reaction is essential for their confirmation.

7.8.3 Choice of medium for isolation and enumeration.

Minerals Modified Glutamate Medium (MMGM) (PHLS 1969) is recommended for the isolation of coliforms from waters and a Durham tube can be included in each bottle or tube of medium for the detection of gas production, if the information concerning gas production is required or considered relevant. This medium has been found to be slightly superior to Lauryl Tryptose Lactose Broth (LTLB) (APHA 1989) especially with chlorinated waters and for samples with coliform counts below 50 per 100 mL of water (PHLS/SCA 1980b). LTLB is, however, an acceptable alternative medium to MMGM.

The use of MacConkey broth is not recommended because of variations in the inhibitory properties of different batches of bile salts.

7.8.4 Practical considerations

The method was originally developed for the detection of coliforms on the basis of acid **and** gas production from lactose in a suitable isolation medium. In the context of the definition given in sections 3.2, 7.7.3 and 7.8.2 the production of acid from lactose serves as the primary indicator for the detection of coliforms. The determination for gas may however be useful in the characterisation of these organisms. As some non-coliform organisms can also produce acid and gas it may be necessary, for further information to be obtained, to subculture to more selective media such as LTLB or Brilliant Green Lactose Bile Broth (BGLBB). These media do not contain pH indicators and fermentation of lactose is demonstrated **only** by the production of gas. Most coliforms produce gas from lactose and can be detected using LTLB or BGLBB. Anaerogenic (non gas producing) strains cannot be detected using these media and subculture to MA is necessary. Subculture from cultures showing acid production in MMGM to MA is likely to give rise to mixed growth whereas subculture to LTLB or BGLBB followed by incubation and subculture to MA will give better selection of coliforms.

7.8.5 Procedure

Prepare the sample as described in section 7.4. Inoculate bottles and tubes of MMGM as described in section 7.6.2.2.

Incubate the bottles and tubes at 37°C and examine for acid (and gas production if this information is required or considered relevant) after 18–24 hours and after a further 24 hours. Retain all positive cultures as they may be needed for repeat subcultures.

Subculture all cultures in MMGM showing acid, with or without gas, to two tubes of either LTLB or BGLBB containing Durham tubes. Make a note of those tubes subcultured from primary cultures in MMGM which did not produce gas.

Incubate one tube at 37°C for 48 hours and the other at 44°C for 24 hours. The production of gas at 37°C can help in the characterisation of coliforms and gas at 44°C the characterisation of *E. coli*.

Examine tubes of LTLB or BGLBB subcultured from anaerogenic cultures in MMGM for growth as indicated by the presence of turbidity. If growth is present, subculture to MA and incubate at 37°C for 18–24 hours. The appearance of coliforms including *E. coli* on this medium is described in section 7.8.7.3.

7.8.6 Confirmation of total coliforms

For each specimen, subculture from at least one tube of LTLB or BGLBB at 37°C to MA and NA. If all primary cultures in MMGM were anaerogenic then subculture from at least one tube of LTLB or BGLBB showing growth.

Incubate the plates of MA and NA at 37°C for 18–24 hours. The colonial morphology of typical coliform organisms is described in section 7.8.7.3. If a pure culture is obtained on NA then perform the oxidase test as described in section 7.8.7.2. If there is any doubt about the purity of the culture then subculture at least one typical coliform colony from MA to NA, incubate at 37°C for 18–24 hours and carry out the oxidase test. Confirmed coliform organisms are oxidase negative.

7.8.7 Confirmation of *E. coli*

Subculture from all MMGM tubes showing acid, with or without gas to TW and incubate at 44°C for 24 hours. The presence of *E. coli* can be demonstrated by growth and production of gas in LTLB or BGLBB and indole in TW (see section 7.8.7.1). Lauryl Tryptose Mannitol Broth with added Tryptophan (PHLS/SCA 1980c) can be a suitable alternative single tube medium for the detection of *E. coli* on the basis of gas and indole production at 44°C. Single tube media may occasionally give false negative results in the indole test and it is essential that a negative indole reaction is repeated by subculture from the culture in MMGM to TW. Anaerogenic indole positive isolates should be tested further for confirmation as *E. coli* after subculture to MA.

7.8.7.1 Indole test

Incubate TW at 44°C for 24 hours and add 0.2–0.3 mL of Kovacs' reagent (Kovacs 1928). Indole production is demonstrated by the rapid appearance of a deep red colour in the upper layer.

7.8.7.2 Oxidase test

Some bacteria found in water may conform to the definition of coliform organisms in most respects, but are able to produce gas from lactose only at temperatures below 37°C. *Aeromonas* species, which occur naturally in water, have an optimum growth temperature in the range 30–35°C but may nevertheless produce acid, with or without gas, from lactose at 37°C. They are of uncertain public health significance and are distinguishable from the coliform group by a positive oxidase reaction. The oxidase test is carried out with pure cultures of lactose-fermenting organisms grown on NA as follows:

Place 2–3 drops of freshly prepared oxidase reagent on to a filter paper in a Petri dish.

With a platinum (not nichrome) wire loop, plastic loop, wooden stick or glass rod, smear some of the growth on the prepared filter paper.

Regard the appearance of a deep blue-purple colour within 10 seconds as a positive reaction.

Commercial test kits for oxidase testing are available and should be used in accordance with manufacturers' instructions.

On each occasion that oxidase reagent is used, conduct control tests with an organism known to give a positive reaction (for example *Ps. aeruginosa*) and one known to give a negative reaction (for example *E. coli*).

7.8.7.3 Growth on MacConkey agar

Colonies of coliform organisms are usually circular in shape, convex or low convex with a smooth surface and an entire edge. They are red but the depth of colour varies considerably and colonial difference cannot be relied upon for differentiation within the group.

7.8.8 Presence-absence tests for total coliform organisms

7.8.8.1 Principle

The presence-absence (P-A) test is a simple modification of the multiple tube procedure, simplified in that it incorporates one large volume of medium (100 mL) instead of a series of tubes of different volumes. The procedure is based on the principle that coliforms should be absent from 100 mL of sample, and a presumptive positive result therefore indicates that coliforms may be present and that appropriate further action must be taken.

Presence-absence test kits for total coliforms are an acceptable method for testing water leaving a treatment works where daily sampling is required. At the time of publication of this Report a report on the use of presence-absence test kits is expected within the DoE (PECD 7/7/424) and at present, advice is that these kits should not be used on more than two sampling occasions in any one week. They may also be useful for non-routine samples taken in distribution systems, for example after a mains repair.

These test kits should be disposed of safely in accordance with documented procedures.

7.8.8.2 Choice of medium

As with the multiple tube method, MMGM is the recommended medium.

7.8.8.3 Procedure

Volumes of 100 mL of double strength medium are decanted into screw-capped bottles of at least 250 mL capacity, with, if necessary, each bottle containing a large inverted Durham tube. The bottles are then sterilized by heating at 115°C for 10 minutes. The caps should be tightened when the cooled bottles are removed from the autoclave. P-A kits may be stored at room temperature in the dark for up to one month.

Volumes of 100 mL of sample are added to each bottle, which, for convenience, can be marked at the 200 mL level. The sample can be poured directly into the bottle, following the procedures given in section 6.4.8. After replacing the cap, the bottle should be inverted to remove any gas present in the Durham tube which may have been added to the bottle. The addition of chlorine-neutralising agents is unnecessary.

7.8.8.4 Incubation and interpretation of results

The bottles should be incubated at 37°C and examined after 18 to 24 hours. The presence of acid should be regarded as a presumptive positive and appropriate action should be taken as necessary. Under these circumstances a 'safety first' principle should be adopted and any test showing the presence of acid only should be regarded as presumptive until confirmatory tests have established otherwise.

If a negative result is obtained after 18–24 hours, the bottles should be incubated for a further 24 hours to obtain a final result. The presence of acid indicates a positive result, and tests should be carried out to confirm the presence of coliforms, *E. coli* and any other organisms as deemed necessary by the microbiologist.

A positive result from a treated water sample should initiate an immediate re-sample for analysis by a quantitative method in a microbiology laboratory.

7.9 Tests for Faecal Streptococci

7.9.1 Introduction

In the United Kingdom faecal streptococci are regarded as secondary indicators of faecal pollution, and the main use of the faecal streptococcus test is to assess the significance of coliform organisms in a sample in the absence of *E. coli*. Occasionally, identification of the species of streptococci present may help to distinguish between human and animal pollution (section 3.5).

7.9.2 Definitions

In the context of these methods, faecal streptococci are Gram-positive cocci which form pairs or chains and possess Lancefield's Group D antigen. They can grow in the presence of bile salts, in concentrations of sodium azide inhibitory to coliform organisms and most other Gram-negative bacteria, and at a temperature of 44° ± 0.25°C. They also hydrolyse aesculin and are catalase-negative. Some species are resistant to heating at 60°C for 30 minutes, grow in nutrient broth containing 6.5% sodium chloride, and to pH 9.6.

Enterococcus faecalis and some related species reduce 2,3,5-triphenyltetrazolium chloride (TTC) to the insoluble red dye formazan. Some species are resistant to heating at 60°C for 30 minutes, to pH 9.6, and can grow in nutrient broth containing 6.5% sodium chloride.

7.9.3 Toxicity of sodium azide

Many of the media described in this section contain sodium azide. As this substance is highly toxic, great care should be taken in preparing these media, especially when dehydrated ingredients in powder form are used. Sodium azide also has the property of forming explosive compounds with metals, especially copper and lead. Waste material containing sodium azide must therefore be discarded into drains with care, preferably through plastic pipes. Azide compounds should be decomposed and rendered safe by an excess of sodium nitrite before disposal.

7.9.4 Count of faecal streptococci by membrane filtration

7.9.4.1 Principle

Isolation of organisms on a membrane placed on the surface of an agar medium containing TTC. Faecal streptococci usually produce pink, maroon or red colonies as a result of the formation of formazan.

7.9.4.2 Definition

In the context of this method, presumptive faecal streptococci usually reduce TTC to insoluble red formazan to produce red, maroon or pink colonies on Membrane Enterococcus Agar (MEA) (Slanetz and Bartley 1957) after incubation at an appropriate temperature. Some strains may produce pale or colourless colonies. Confirmation is on the basis of aesculin hydrolysis on Bile Aesculin Agar (BAA) or Kanamycin Aesculin Azide Agar (KAAA) incubated at 44°C for 18 hours. Further tests as detailed in section 7.9.6 may also be carried out.

7.9.4.3 Choice of medium for isolation and enumeration

The medium recommended for the isolation and enumeration of faecal streptococci is MEA. Complete dehydrated medium is available commercially and should be prepared exactly as described by the manufacturer, with special emphasis on the avoidance of overheating. If the medium is orange or pink when cooled to 50°C, then consideration should be given as to whether or not it should be discarded. Proper QC validation is particularly important in these cases.

7.9.4.4 Practical considerations

Growth of faecal streptococci on MEA is better at 37°C although some organisms resembling faecal streptococci may grow which cannot be confirmed as such. Selectivity is better at 44°C although lower counts of faecal streptococci may be obtained. It is therefore recommended that membranes from samples of potable water are incubated at 37°C and from untreated waters at 44°C. Although an incubation temperature of 45°C is more selective than 44°C, inhibition of the more temperature-sensitive faecal streptococci may result from a rise even of as little as less than 0.5°C.

7.9.4.5 Procedure

Prepare and filter the sample as described in section 7.5, making any necessary dilutions. Place the membrane on the surface of a well-dried plate of MEA.

For potable waters, incubate at 37°C for 48 hours and for untreated waters incubate at 37°C for 4 hours followed by 44°C for 44 hours.

After incubation, count all red, maroon or pink colonies which are smooth and convex in shape. These are presumptive faecal streptococci. Some types of faecal streptococci may produce pale or colourless colonies. Colonial size is variable but is usually not less than 0.5 mm. *Bacillus* species may produce pink colonies but these are often rough, flat and sometimes spreading. Depending on (i) whether the aim is to estimate the count or demonstrate the presence/absence of organisms, and (ii) the number of colonies present on the membrane, subculture a suitable number of red, maroon, pink or colourless colonies from the membrane. See section 6.3.3.

7.9.4.6 *Confirmation of faecal streptococci*

Subculture to BAA and incubate at 44°C for 18 hours. Faecal streptococci should produce discrete colonies surrounded by a brown or black halo from aesculin hydrolysis. The development of this colour is usually evident within a few hours and will give rapid confirmation. *Bacillus* species may produce some discolouration around the original inoculum site but should not develop discrete colonies. KAAA may be used instead of BAA. The use of Glucose Azide Broth (GAB) at 44°C is not recommended as a confirmatory test as false positive results may be obtained with other Gram-positive organisms. Also the transfer of membranes from the primary isolation medium to KAAA is not recommended because of the spreading of the black colour to aesculin-negative colonies.

7.9.5 Count of faecal streptococci by multiple tube

7.9.5.1 *Principle*

Culture of the sample in a liquid enrichment medium containing sodium azide, glucose and bromocresol purple as indicator. Growth and acid production indicate a positive result and therefore the presumptive presence of faecal streptococci. Subculture to confirmatory media to demonstrate growth at 44°C and hydrolysis of aesculin.

7.9.5.2 *Choice of medium*

GAB (Mallman and Seligmann 1950) is recommended as the liquid enrichment medium.

7.9.5.3 *Procedure*

After preparation of the sample and making any necessary dilutions as described in section 7.6.2, inoculate the test volumes into tubes of single or double strength GAB as described in section 7.6.2.1.

Incubate the tubes at 37°C and examine after 24 and 48 hours. After 48 hours regard any trace of acidity as a positive reaction.

7.9.5.4 *Confirmatory test*

From each tube giving a positive reaction, place one drop (about 0.02 mL) or loopful of growth onto the surface of a BAA plate and incubate at 44°C. Development of a black or brown colour in or around the inoculum within a few hours confirms the presence of faecal streptococci. If the inocula are well spaced out, numerous cultures can be tested on one plate. Incubate for up to 18 hours.

From the number of tubes giving a positive confirmatory reaction, calculate the MPN of faecal streptococci in 100 mL of the sample by reference to the tables in Appendix C.

The catalase test is used to distinguish streptococci, which are catalase-negative, from other Gram-positive cocci. Since this test can only be done with pure cultures, its application in the MPN method requires subculture to solid media.

7.9.6 Additional confirmatory and differential tests for faecal streptococci

Although the possession of Lancefield's Group D antigen is referred to in the definition, serological methods of confirmation present many practical difficulties and their use is therefore not recommended in routine water examination.

Tolerance of 40% bile is also characteristic of faecal streptococci, but in practice this test is not usually necessary and growth on MA can be used instead. Further tests with subcultures may be done if necessary, partly as an aid to species differentiation. However, full identification depends on the demonstration of biochemical and other characteristics as described by Cowan (1993) or by means of one of the multi-test differential systems now available.

Enterococcus species are differentiated from other streptococci by their ability to grow in Nutrient Broth (NB) containing 6.5% sodium chloride and in Glucose Phenolphthalein Broth (GPB) (Clarke 1953) at pH 9.6.

7.9.6.1 Catalase test

Emulsify some of the culture to be tested in a few drops (about 0.1 mL) of quarter-strength Ringer's solution in a small screw-capped bottle. Add 0.05 mL of 3% hydrogen peroxide solution and replace the cap. The appearance of bubbles (of oxygen) indicates catalase activity. An alternative method is to add the hydrogen peroxide to an overnight culture of the organisms in a tube of either NA or NB. The test should preferably not be performed on a slide because of the risk of aerosol formation.

7.9.6.2 Bile tolerance

Subculture to a plate or tube of Bile Agar (BiA) (40%) and incubate at 37°C for 24–48 hours. Growth on this medium indicates tolerance of bile salts. Alternatively, use MA to show growth in the presence of bile salts. Faecal streptococci form small deep red colonies on this medium.

7.9.6.3 Heat resistance

Transfer 1 mL of a 24 hour broth culture to a small test tube, and place in a water bath at 60°C for 30 minutes. Cool the tube rapidly and incubate at 37°C for 24 hours. Subculture to a Blood Agar (BlA) plate or other non-selective medium, incubate and examine for growth.

7.9.6.4 Growth at pH 9.6

Inoculate a tube of GPB and incubate at 37°C for 24 hours. Tolerance of pH 9.6 is indicated by heavy growth and de-colourization of the medium.

7.9.6.5 Salt tolerance

Inoculate a tube of NB containing 6.5% of sodium chloride (Salt Broth) and incubate at 37°C for 24–48 hours. Examine for growth.

7.10 Tests for Sulphite-reducing Clostridia and *Clostridium perfringens*

7.10.1 Introduction

The tests for sulphite-reducing clostridia play only a subsidiary role in water examination. These organisms form spores which are resistant to heating compared with vegetative cells and advantage is taken of this for the detection of clostridia in water. *Cl. perfringens*, an important member of this group, is associated with faecal contamination. If found at a time when other faecal indicator organisms are no longer detectable, it can indicate remote or intermittent pollution.

7.10.2 Definitions

In the context of the methods, sulphite-reducing clostridia are Gram-positive anaerobic spore-forming rods which reduce sulphites to sulphides. *Cl. perfringens* forms a stormy clot in Litmus Milk Medium (LMM) or Crossley's Milk Medium (CMM) (Crossley 1941).

7.10.3 Count of sulphite-reducing clostridia and *Clostridium perfringens* by membrane filtration (Tentative)

7.10.3.1 Principle

This method has not been fully validated and several media are described. After preliminary treatment to destroy vegetative bacteria, filtration of a volume of the sample through a membrane, followed by anaerobic incubation on a sulphite-containing agar medium. Count all black colonies that develop. Subculture to a confirmatory medium to demonstrate the presence of *Cl. perfringens*.

7.10.3.2 Choice of media

Several variations of sulphite-containing media and culture methods for anaerobic growth are available for the detection of sulphite reducing clostridia. The following media are recommended on a tentative basis:

(i) Tryptose-Sulphite-Cycloserine Agar (TSCA) containing egg yolk (Harmon, Kautter and Peeler 1971) or without egg yolk (Hauschild and Hillsheimer 1974, Sartory 1986);

(ii) Shahidi—Ferguson Perfringens Agar (SFPA) (Shahidi and Ferguson 1971);

(iii) Oleandomycin—Polymyxin—Sulphadiazine Perfringens Agar (OPSPA) (Handford 1974).

These are available as commercially dehydrated media and should be assessed for recovery in parallel with Differential Re-inforced Clostridial Medium (DRCM) (section 7.10.4) (Gibbs and Freame 1965).

7.10.3.3 Practical consideration

Funnels to be used for clostridia should be sterilized by autoclaving as spores may not be inactivated by boiling.

7.10.3.4 Procedure

Heat the sample to 75°C in a water bath and maintain at this temperature for 10 minutes. The time needed to reach this temperature can be determined with a similar bottle containing the same volume of water and a thermometer.

After heat treatment, prepare and filter the sample as described in section 7.5, making any necessary dilutions.

Place the membrane face-upwards or inverted on the agar and carefully pour 15–20 mL of agar, pre-cooled to 50°C over the membrane and allow to set. The agar should be poured in such a manner to avoid the washing-off of bacterial spores. Overlaying the agar is optional but may give better results; this should be assessed in individual laboratories.

Incubate in an anaerobic jar or cabinet at 37°C. Examine the plates after 24 and 48 hours and count all black colonies.

7.10.3.5 Confirmation of Cl. perfringens

Using a straight wire, subculture each colony to be tested (see 6.3.3) into a tube or bottle of freshly steamed and cooled LMM or CMM. For colonies on inverted membranes it will be necessary to stab through the membrane. Incubate the milk media as described in section 7.10.4.4 below and regard a stormy clot reaction as confirmation of *Cl. perfringens*. Black colonies should also be subcultured to two nutrient agar plates, one to be incubated aerobically at 37°C for 48 hours to exclude *Bacillus* species, which are facultative anaerobes, and the other anaerobically as a check for purity.

7.10.4 Count of sulphite-reducing clostridia and *Clostridium perfringens* by multiple tube

7.10.4.1 Principle

After preliminary treatment to destroy vegetative bacteria, test volumes of the sample are cultured in bottles of a liquid medium containing sodium sulphite and iron(III) citrate. All bottles showing blackening are subcultured to a confirmatory medium.

7.10.4.2 Choice of medium

DRCM is recommended for isolation. It should be distributed in screw-capped bottles instead of tubes, with sufficient depth of medium to ensure adequate anaerobiosis during incubation. LMM and CMM are recommended for confirmation.

7.10.4.3 *Procedure*

Heat the sample to 75°C in a water bath and maintain at this temperature for 10 minutes. After heat treatment, make any dilutions necessary as described in section 7.4.2.

Add the 50 mL and 10 mL test volumes to equal volumes of double-strength medium and the 1 mL volumes to 25 mL of single-strength medium.

Top-up with single-strength medium so as to leave a minimum amount of air space.

A considerable amount of gas may be produced in this test especially for the 50 mL volumes; the use of stout glass or plastic bottles is advisable. Each of these should be placed inside its own plastic bag in order to contain any breakage or spillage during incubation.

Incubate the bottles at 37°C for 48 hours. A positive reaction is shown by blackening of the medium caused by reduction of sulphite and the precipitation of iron(II) sulphide. A count made at this stage represents the MPN of spores of sulphite-reducing clostridia.

7.10.4.4 *Confirmation of Cl. perfringens*

Transfer a loopful (0.02 mL) from each bottle showing a positive reaction to a tube of LMM or CMM which has been freshly steamed, to expel dissolved oxygen, and cooled. Growth is improved in the litmus milk by adding to each tube a nail or a short length of iron wire sterilized by heating to redness immediately before inoculation. Alternatively, the tubes may be incubated in an anaerobic jar or cabinet.

Incubate at 37°C for 48 hours. Tubes containing *Cl. perfringens* will show a 'stormy clot' reaction, in which, as a result of lactose fermentation, the milk is acidified and coagulated, and the clot is disrupted by gas and often blown up the tube.

From the number of tubes showing a positive reaction, calculate the MPN of *Cl. perfringens* spores in 100 mL of the sample by reference to the appropriate tables in Appendix C.

7.11 Colony Count

7.11.1 Introduction

The colony count of heterotrophic bacteria may be enumerated by the pour or spread plate methods using Yeast Extract Agar (YEA) or R2A Agar (Reasoner and Geldreich 1985). Plates are incubated at 22°C and 37°C as detailed in section 7.11.6. The significance of the results obtained is discussed in section 3.6.1.

The most useful application of the colony count is to detect change, especially sudden change, in the bacterial content of waters. Comparisons between results for a given water can only be made if the same method is used each time. The method should be stated in the report. Colony counts at 37°C are useful to assess the quality of relatively unpolluted groundwaters and can provide an early indication of more serious pollution.

7.11.2 Definition

In the context of these methods, colony forming units (CFU) comprise bacteria, yeasts and moulds capable of growth under the conditions specified.

7.11.3 Choice of medium

The pour plate method using YEA is used in the UK for compliance monitoring.

For special investigations such as complaints of taste or odour it may be advantageous to use R2A Agar, or the spread plate method using either medium. R2A Agar is a low nutrient medium which enhances the recovery of disinfectant-damaged organisms and those organisms with a low nutrient requirement which may be inhibited by the higher concentrations of nutrient in YEA. The spread plate method will give better recovery than the pour plate method as organisms are not heat shocked by the addition of molten agar. Furthermore, subculture from colonies is facilitated by use of the spread plate method.

7.11.4 Colony count by the pour plate method

7.11.4.1 Principle

Mixing of test volumes of the water sample with molten YEA or R2A Agar in Petri dishes. Incubation under the conditions specified and a count of the colonies that develop.

7.11.4.2 Procedure

Prepare the sample as described in section 7.4, making any necessary dilutions. Starting with the most dilute sample solution, pipette a 1 mL volume of each dilution (and the original sample) into separate, empty sterile Petri dishes.

Pour about 20 mL of molten agar, pre-cooled to 45–50°C, into each Petri dish within 20 minutes of dispensing the 1 mL sample volumes and cover the dish with a lid. If the medium has been cooled in a waterbath, wipe off excess water from the container prior to pouring so as to avoid contamination from the side of the container. Mix the sample and medium by rapid but gentle to and fro circular movements for approximately 10 seconds. Lift the lid slightly during mixing so that no solution adheres to the lid and keep the Petri dish flat on the bench. Allow the solution to solidify and incubate in an inverted position.

7.11.5 Colony count by the spread plate method

7.11.5.1 Principle

Spreading of test volumes of sample onto the surface of a pre-dried agar medium in Petri dishes. Incubation under specified conditions and a count of the colonies that develop.

7.11.5.2 Procedure

Prepare the sample as described in section 7.4, making any necessary dilutions. Pipette a volume of 0.1 mL on the surface of a pre-dried agar medium and spread with a sterile bent glass or plastic rod. Alternatively, spread by rotating the dish.

Allow thc inoculum to soak into the agar, and incubate in the inverted position.

7.11.6 Incubation and counting of colonies

For YEA, incubate one set of plates at 22°C ± 1°C for 72 ± 3 hours and another set at 37°C for 24 ± 3 hours. Plates incubated at 37°C may be incubated for up to 48 hours if further information is required.

For R2A Agar, incubate one set of plates at 22°C ± 1°C for 5–7 days and another set at 30°C for 3 days.

Count the colonies as soon as the plates are removed from the incubator. Count the colonies on the plate containing the original sample. If the count is greater than 300 then count the colonies from a diluted sample containing between 30–300 colonies. If all plates carry more than 300 colonies record this as greater than 300 at the highest dilution plated and express the sample count in terms of this. Do not use the term 'too numerous to count' without further qualification (see section 6.3.5).

Spreading colonies can interfere with counts and the following gives guidance on what should be counted a single CFU.

(i) A chain of colonies that appears to be caused by disintegration of a clump of organisms.

(ii) A spreading growth that develops as a film at the bottom of the Petri dish.

(iii) A colony that forms in a film of water at the edge or over the surface of the agar.

If the colonies cannot be counted immediately then keep the plates at 4°C for no longer than 24 hours.

7.12 Acridine Orange Direct Count

7.12.1 Introduction

The acridine orange direct count (AODC) produces a count of all viable and non-viable microbial cells in the sample. Consequently the AODC may be higher than the colony count (section 7.11) as the latter method will only give a count of viable micro-organisms. The main value of the AODC in water microbiology is to give a rapid estimate of the total microbial biomass, living and dead. All reagents should be free of bacteria and blanks should be used to check on background levels.

7.12.2 Definition

The AODC is a count of all viable and non-viable bacteria, yeasts and moulds in a water sample (Hobbie, Daley and Jasper 1977, APHA 1989).

7.12.3 Principle

Fixing the sample for storage, staining with acridine orange followed by vacuum filtration on to a non-fluorescing polycarbonate membrane and enumeration using an epifluorescence microscope.

7.12.4 Apparatus

The following apparatus can be used.

Blender or vortex mixer.
Filtration units suitable for use with 25 mm diameter membrane filters.
Test tubes, glass screwcapped, 13 × 125 mm.
Syringes, 3 mL disposable with disposable syringe filters of 0.2 µm pore size.
Membrane filters, non-fluorescent, polycarbonate 25 mm diameter, 0.2 µm pore size, also cellulose 25 mm diameter 5 µm pore size.

7.12.5 Microscopy

A suitable microscope with appropriate filters.

Details of the practical aspects of fluorescent microscopy can be found in 'Immunofluorescence: antigen-detection techniques in diagnostic microbiology' (PHLS 1992).

7.12.6 Reagents

Phosphate buffer. Dissolve 13.6 g potassium dihydrogen phosphate (KH_2PO_4) in water and dilute to 1 litre. Adjust the pH to 7.2 and filter through a 0.2 µm membrane filter.

Fixative solution. A 5.0% m/V glutaraldehyde solution in phosphate buffer. Prepare daily.

Staining solution. An 0.1% m/V acridine orange solution in phosphate buffer.

Immersion oil, low fluorescing.

7.12.7 Procedure

7.12.7.1 Fixation

Prepare the sample as described in section 7.4. Transfer 9 volumes of the sample to a test tube containing 1 volume of fixative solution. The fixed sample can be stored for up to 3 weeks at 4°C.

7.12.7.2 Staining

Prior to staining mix the fixed sample using a blender or vortex mixer and if necessary, make ten-fold dilutions in phosphate buffer, pH 7.2. Place 1 volume of the fixed sample in a small clean container and add 1 volume of acridine orange solution with a sterile disposable syringe filter. Leave for 2 minutes and then transfer to a non-fluorescent polycarbonate filter placed on a cellulose membrane filter in a filter holder. Alternatively, mix the fixed sample and acridine orange on a non-fluorescent filter, leave for 2 minutes and add approximately 3 mL of filtered phosphate buffer to give an even distribution of cells on the membrane.

For both methods apply a vacuum (about 13 kPa, 100 mm of mercury), wash with 2 mL phosphate buffer and refilter. Remove the filter using forceps and allow to air dry for 1–2 minutes.

7.12.7.3 Microscopic examination

Place a drop of immersion oil on a clean glass microscope slide and then place the filter on the immersion oil. Add another drop of immersion oil to the surface of the filter and place a coverslip over the filter.

Examine using a x100 oil immersion lens. Carry out a preliminary examination of at least 10 random fields of view to ensure that distribution of cells is uniform and that cells can be counted (preferably 10–50 per field). Count the number of cells observed in at least 20 squares, preferably 50, of a calibrated counting graticule.

7.12.7.4 Calculation of the acridine orange direct count

Calculate the average number of cells per square and the area of the filter through which the sample was filtered using the specifications of the filter unit. Calculate the number of cells per mL thus:-

$$\text{Total cells per mL} = \frac{abc}{d}$$

where a = average number of cells per square
b = total number of squares per filter
c = the dilution factor, if applicable
d = the sample volume, in mL

7.13 Test for *Pseudomonas aeruginosa*

7.13.1 Introduction

Pseudomonas aeruginosa should not be used as an indicator of faecal pollution. However, in the manufacture of food, drink and pharmaceutical products and in hospitals, it is desirable that the water should be free from this organism. In these circumstances, and for the examination of bottled waters, the following methods of detection and enumeration may be used.

7.13.2 Definition

In the context of the methods, *Ps. aeruginosa* is an aerobic, Gram-negative non-sporing bacillus possessing oxidase and catalase, grows at 42°C but not at 4°C and shows oxidative metabolism in the test of Hugh and Leifson (1953). It reduces nitrates and nitrites, produces ammonia from the breakdown of acetamide, liquifies gelatin and hydrolyses casein but not starch. One of the most important characters is the production of the blue-green pigment pyocyanin, and/or the fluorescent pigment fluorescein.

7.13.3 Count of *Pseudomonas aeruginosa* by membrane filtration

7.13.3.1 Principle

Filtration of a test volume of the sample through a membrane. Incubation on an absorbent pad saturated with broth containing ethanol and cetrimide, followed by a count of the pigment-producing colonies. Alternatively, a solidified version of the medium containing selective agents may be used. Subculture to a confirmatory medium to show growth at 42°C with casein hydrolysis and the production of pyocyanin and/or fluorescein.

7.13.3.2 Choice of medium

Modified King's A Broth (MKB) (Medium 19 of Drake 1966) is recommended for isolation or a solidified version supplemented with cephaloridine, fucidin and cetrimide

(Mead and Adams 1977). This medium, referred to as Pseudomonas Agar (PA) is available commercially. Milk Agar with Cetrimide (MAC) (Brown and Foster 1970) is recommended as the confirmatory medium.

7.13.3.3 Procedure

After preparing the sample and making any necessary dilutions (see section 7.4) filter the test volumes as in section 7.5 and place each membrane on a pad saturated with MKB in a Petri dish or on the surface of a well-dried plate of PA. Incubate at 37°C for 48 hours (in a closed container if MKB is used). Count all colonies which produce a green, blue or reddish-brown pigment and those which fluoresce under UV light. To detect fluorescence, it is advisable to use within a darkened room a UV light source (wavelength 350 ± 20 nm) with a dark (Woods' glass) filter to eliminate visible light. Exposure of the colonies to air for a short time enhances the pigmentation and thus facilitates counting.

7.13.3.4 Confirmation of Pseudomonas aeruginosa

Regard those colonies which clearly produce pyocyanin on the membrane as *Ps. aeruginosa*. To confirm other colonies, subculture each colony to be tested (see 6.3.3) to MAC and incubate at 42°C as described for the multiple tube method (section 7.13.4.4).

7.13.4 Count of *Pseudomonas aeruginosa* by multiple tube

For most waters where it is necessary to enumerate *Ps. aeruginosa*, membrane filtration is likely to be more suitable than the multiple tube method. The latter should be used only when membrane filtration is difficult or impossible because of the presence of excess particulate matter. Since such waters are likely to be polluted, the use of test volumes greater than 10 mL should rarely be necessary.

7.13.4.1 Principle

Culture of test volumes of the sample in a liquid medium containing asparagine and ethanol. Examination for growth and pigment formation at 38–39°C. This temperature is used to inhibit the growth of other pseudomonads which are common in water and soil and which grow better at a lower temperature. Subculture to a confirmatory medium to show growth at 42°C with pigment formation, casein hydrolysis and cetrimide resistance.

7.13.4.2 Choice of medium

Asparagine Broth and Ethanol (ABE) (Drake 1966) is recommended for isolation; each test volume of the sample is usually added to four times its volume of medium. Alternatively, for test volumes of 1 mL, Repli-dishes instead of tubes may be used with the more concentrated form of the medium. Repli-dishes are Petri dishes which are divided into 25 square compartments each of 5 mL capacity. For confirmation, MAC is recommended.

7.13.4.3 Procedure

After preparing the sample and making any necessary dilutions as described in section 7.6.2, add 10 mL test volumes to 40 mL of medium in screw-capped bottles, and 1 mL volumes to 4 mL of medium in tubes or to 1 mL of concentrated medium in Repli-dishes.

Incubate at 38–39°C and examine after 48 hours for growth and fluorescence. This is best carried out in a darkened room under UV light of wavelength 350 ± 20 nm; it is important to exclude visible light from the UV source by means of a dark (Woods' glass) filter. Examine after 96 hours for green, blue or red pigments in the medium. Regard growth and pigment production as presumptive evidence of the presence of *Ps. aeruginosa*.

7.13.4.4 Confirmation of Ps. aeruginosa

Subculture from each tube, bottle or Repli-dish showing fluorescence or pigmentation to a plate of MAC.

Incubate at 42°C for 24 hours. Examine for growth, pigment formation and for casein hydrolysis which is shown by clearing of the medium around the colonies. *Ps. aeruginosa* grows at 42°C, hydrolyses casein and produces pyocyanin and/or fluorescein. Occasionally, non-pigmented variants of *Ps. aeruginosa* may occur.

From the number of tubes giving a positive reaction, calculate the MPN of *Ps. aeruginosa* per 100 mL of the sample by reference to the appropriate tables in Appendix C.

If only a qualitative result is required, report *Ps. aeruginosa* present or absent from the volume of water examined.

7.14 Test for Motile *Aeromonas* Species (Tentative)

7.14.1 Introduction

The taxonomy of the genus *Aeromonas* is still the subject of considerable research. At least 13 species can be identified by molecular genetic techniques. Using simple biochemical tests it is not possible to identify these species accurately but most isolates can be broadly divided into three groups corresponding to *A. hydrophila*, *A. caviae* and *A. veronii* biotype sobria. At present all the aeromonads may be considered equally significant and so for practical purposes it is not necessary for the water microbiologist to identify aeromonads beyond the level of the genus.

7.14.2 Definition

For the purposes of the method strains are identified as *Aeromonas* if they are oxidase-positive, resistant to ampicillin (10 μgmL^{-1}) and ferment dextrin and mannitol. In addition, they are Gram-negative, fermentative in the test of Hugh and Leifson (1953), able to grow in 1% tryptone water containing no added sodium chloride but not in tryptone water containing 6% sodium chloride, resistant to 50 μgmL^{-1} 0/129 phosphate (2,4-diamino-6,7-diisopropyl pteridine phosphate) and able to hydrolyse arginine in the test of Thornley (1960).

7.14.3 Count of *Aeromonas* by membrane filtration

7.14.3.1 Principle

Filtration of the test volume of sample through a membrane and isolation of the organisms by placing the membrane on an agar medium containing ampicillin as a selective agent, dextrin as a fermentable carbohydrate and bromothymol blue as an indicator of acidity. Subculture of presumptive isolates to confirm identity by the oxidase test and fermentation of mannitol.

7.14.3.2 Choice of medium

A variety of media has been developed for the isolation of aeromonads in recent years. Many of them contain ampicillin as the selective agent. Ampicillin-dextrin agar (ADA) of Havelaar, During and Versteeght (1987) is recommended. This is the same medium of Rippey and Cabelli (1979) except that it has been made more specific by the substitution of dextrin for trehalose. Some *Vibrio* species can also grow on ADA with similar reactions to *Aeromonas*. If the samples are likely to contain strains of *Vibrio*, as would be the case for estuarine waters, then ADA can be made more selective by the inclusion of the O/129 phosphate at a concentration of 50 mgL^{-1}.

7.14.3.3 Procedure

After preparing the sample and making any required dilutions as described in section 7.4, filter the test volume as in section 7.5 and place each membrane on the agar surface in a Petri dish containing ADA. Incubate aerobically at 29–31°C for 24 hours. Count all yellow colonies.

7.14.3.4 Confirmation

Subculture all or a representative number of colonies to NA. After overnight incubation at 30°C test colonies for oxidase. Oxidase-positive colonies are presumptive *Aeromonas*.

However, some vibrios may also give similar reactions. The identification may be confirmed by testing in order to show;

(i) fermentation of glucose in the test of Hugh and Leifson (1953),

(ii) fermentation of mannitol,

(iii) hydrolysis of arginine in the test of Thornley (1960),

(iv) growth in 1% tryptone water without added sodium chloride, and

(v) inhibition of growth in 1% tryptone water containing 6% sodium chloride.

All tests are incubated for 24 hours at 30°C.

The single tube confirmation medium of Kaper et al (1979) or commercial identification kits may also be used for identification, although some less halophilic *Vibrio* species, particularly *V. fluvialis* and *V. furnissii* may still be mis-identified as aeromonads.

Chapter 8: Examination for Pathogenic Organisms

8.1 Introduction

The direct search for pathogenic bacteria has no place in the routine microbiological examination of water supplies. There may be occasions, however, when investigations for faecal pathogens will be necessary. Technical difficulties are considerable in that any pathogens present in the water are likely to be greatly outnumbered by normal faecal flora or by putrefactive organisms derived from sewage. The detection of particular pathogens therefore necessitates the examination of large volumes of water by means of concentration techniques and selective enrichment media. This should only be undertaken in a containment level 3 laboratory by trained and experienced staff under accepted codes of safe practice (NWC 1983, HSAC 1991a, 1991b).

All the methods in this chapter are to be regarded as tentative as they have not been fully validated and subject to formal inter-laboratory evaluation. It certain cases it may be possible to adapt the methods to make a quantitative estimate of the number of organisms present.

8.2 Concentration Methods

The most effective way of examining large volumes of water for pathogenic bacteria is by membrane filtration. However, because membranes readily become blocked by suspended solids, this method is applicable only to waters of low turbidity. In some cases it may be necessary to filter the water in small amounts and to replace the membranes when they become blocked.

When filtration through membranes is impracticable, a suitable filter-aid, for example diatomaceous earth should be used which should be of a sufficient particle size to retain most of the bacteria without causing a blockage of the filter. In addition to concentrating the number of organisms, the method also separates bacteria from any toxic agents which may be present in solution.

8.3 Membrane Filtration

8.3.1 Qualitative examination

Carry out the filtration as in section 7.5. For volumes larger than 100 mL the use of 500 mL funnels is recommended. After filtration place the membrane in a suitable enrichment medium for incubation.

8.3.2 Quantitative examination

Filter separate volumes of the sample through separate membranes. One × 500 mL volume and five × 100 mL volumes will give a measurable MPN count if there are less than 18 organisms per litre, (that is, at least one of the volumes shows no growth). If all volumes show growth, the estimate will be >18 per litre.

After filtration, place each membrane in a suitable enrichment medium for incubation.

8.4 Use of Filter-aid

The usual membrane filtration apparatus may be used, but with a sterile absorbent pad in place of the membrane to act as a supporting base for the filter-aid. A small amount of sterile aqueous suspension of the filter-aid (see Appendix B) is first filtered to form an initial layer on the absorbent pad. Further suspension is then mixed with the sample and filtered. By this means the filter-aid, intimately mixed with the bacteria and debris from the water, is continuously deposited on the initial layer. This should retain more than 90 per cent of the bacteria present in the sample, although for dirty waters, greater proportions of filter-aid may be required.

8.4.1 Qualitative examination

Place a sterile absorbent pad in the membrane filtration apparatus. Take a bottle containing 15 mL of sterile filter-aid suspension, shake well and pour into a funnel. Apply suction and filter to form an initial layer. To each litre of sample, add 15 mL of well-shaken filter-aid. Mix well and filter the suspension in 100 mL aliquots. Mix well to re-suspend the filter-aid before each portion of the sample is added to the funnel.

After filtration is complete, remove the funnel carefully and transfer the absorbent pad with the layer of filter-aid to a pre-enrichment medium.

With the same medium, rinse any filter-aid adhering to the funnel into the culture vessel.

8.4.2 Quantitative examination

Filter a measured volume of sample as above. Add the pad with the filter-aid to 100 mL of pre-enrichment medium in a sterile flask or wide-mouthed container. Wash off any adhering filter-aid into the container with the medium from the flask.

Mix well, and for the MPN estimation, pipette 5 × 10 mL volumes into sterile tubes or wide-mouthed containers whilst keeping the suspension well mixed. Keep the remaining 50 mL as a separate culture. If high counts are expected, pipette from the 50 mL volume, 5 × 1 mL, 5 × 0.1 mL volumes into separate tubes, each containing 10 mL of the pre-enrichment medium.

Incubate at the temperature and time specified for the particular organism being sought.

Subculture from the pre-enrichment cultures to selective enrichment media and incubate again at the temperature and time specified.

Subculture to selective solid media to permit recognition of the particular pathogen.

After confirmation, determine the MPN of the pathogen in the original volume examined.

8.5 Sewer—Swab Technique

A useful qualitative method for the detection of pathogenic bacteria in polluted waters is the swab technique described by Moore (1948, 1950), Moore, Perry and Chard (1952). Gauze swabs (SCA 1983) suitably anchored by wire or other means, are immersed in the flow of water for several days. Organisms are entrapped and thus concentrated in the swab, which can then be treated in the same manner as the filter-aided concentrate. In addition, with heavily polluted waters, the liquid from the swab, and/or dilutions prepared from it, can be plated out directly on to agar media, as well as into enrichment medium.

8.6 *Campylobacter*

8.6.1 Introduction

The genus *Campylobacter* comprises two main groups according to growth temperature. The first group will grow at 42°C and are referred to as thermophilic campylobacters. The second group will grow at 25°C but not at 42°C. Thermophilic campylobacters include the important human pathogens *C. jejuni* and *C. coli* and details of methods for their isolation are given below.

8.6.2 Principle

Isolation of organisms on a membrane followed by enrichment and subculture to a selective agar. The broth and agar are both incubated micro-aerobically at 42°C. Confirmation of thermophilic campylobacters is by cell morphology, Gram stain and oxidase test.

8.6.3 Definition

Bacteria in the genus *Campylobacter* are Gram-negative spirally shaped rods which often resemble gulls' wings. They are motile by means of a single polar flagellum giving a characteristic darting motility. Members of the genus are micro-aerobic, requiring an oxygen tension of around 5% and a carbon dioxide tension of around 10%, are oxidase-positive, reduce nitrates but do not produce acid from carbohydrates.

8.6.4 Choice of media for isolation

The Preston Enrichment Broth, (PEB) (Bolton and Robertson 1982, Bolton et al 1983), Preston Selective Agar (PSA) (Bolton and Robertson 1982), and the blood-free modification of the selective agar (Bolton, Hutchinson and Coates 1986) are recommended for the isolation of thermophilic campylobacters from water. The agar media can be obtained commercially as dehydrated products.

8.6.5 Practical considerations

Membrane filtration is recommended for the concentration of campylobacters from water. Concentration by certain filter aids is not effective in removing all organisms of this type from the sample and in this instance is not recommended.

Campylobacters will only grow in a micro-aerobic environment. For enrichment cultures it is necessary to include the micro-aerobic supplement described by George et al (1978). The volume of enrichment broth used should be adjusted so that any container used is completely filled after inoculation. The lid should be sealed tightly prior to incubation.

Selective agars should be incubated in a suitable micro-aerobic atmosphere containing a mixture of 5% oxygen, 10% carbon dioxide and 85% nitrogen. This can be obtained by using commercially available gas generating kits. Since hydrogen is produced, jars should not be opened adjacent to a source of ignition.

8.6.6 Procedure

8.6.6.1 Membrane filtration

Filter the sample as described in section 7.5, using a membrane of pore size 0.2 µm. It may be necessary to use several membranes for each sample if the water is turbid.

8.6.6.2 Enrichment

Place the membrane(s) in a suitable screw-capped container filled almost completely with 50–100 mL of PEB. Seal tightly and incubate at 42°C for 48 hours.

8.6.6.3 Isolation and identification

Subculture from the enrichment broth to PSA or the blood-free modification of this medium. Incubate in a gas jar containing a micro-aerobic gas generating kit at 42°C for 48 hours. Examine the plates for characteristic colonies, using a hand lens if necessary. *Campylobacter* colonies vary in size but are typically flat and watery with an entire edge and may be circular or ovoid. They have a tendency to spread along the lines of inoculation. Carry out an oxidase test on suspect colonies. Variation in colonial morphology is not uncommon so it is advisable to test several colonies for oxidase production. Carry out a Gram-stain on oxidase-positive colonies using 1:20 carbol fuchsin as the counterstain because campylobacters stain weakly with neutral red. Campylobacters produce Gram-negative curved rods which resemble gulls' wings, S-shapes or short spirals. For most practical purposes colonial and cellular morphology and the oxidase test are adequate for identification. If required, species and sub-species differentiation can be carried out using the bio-typing scheme proposed by Skirrow and Benjamin (1980) and described in detail by Lander and Gill (1985).

8.7 *Escherichia coli* 0157:H7

8.7.1 Introduction

E. coli 0157:H7 is a recognised cause of illness characterised by bloody diarrhoea, severe abdominal pain and haemolytic uraemic syndrome sometimes followed by renal failure. The organism is common in cattle and outbreaks related to beef and milk have been recorded. Isolation is based on enrichment in modified buffered peptone water (MBPW) followed by sub-culture to modified sorbitol MacConkey agar (MSMA) and selection of non-sorbitol fermenting colonies. The method described was developed for milk analysis and has not been fully tested and validated for water.

8.7.2 Principle

The method described is suitable for detecting *E. coli* 0157:H7 from water and related materials by membrane filtration and isolation into an enrichment medium. Enrichment broths are inoculated onto MSMA and non-sorbitol fermenting colonies subcultured for biochemical and serological testing.

8.7.3 Definition

E. coli is a Gram-negative, oxidase negative, non sporing bacillus. Most strains are thermotolerant and can grow at 44°C but *E. coli* 0157:H7 grows only poorly at this temperature in non-selective media and will not grow above 41°C in selective media. It fails to ferment sorbitol at 37°C and MSMA can therefore be used as a screen for the differentiation of 0157:H7 from other *E. coli* strains.

8.7.4 Procedure

8.7.4.1 Membrane filtration

Filter suitable volumes of the sample as described in section 7.5 using membranes of nominal pore size 0.45 µm. It may be necessary to use several membranes with very turbid waters.

8.7.4.2 Enrichment

Place membranes into universal containers with 20 mL of the enrichment broth (MBPW). Incubate at 37°C for 24 hours.

8.7.4.3 Isolation and identification

Subculture the enrichment broth onto MSMA and incubate for 24 ± 1 hours at 37°C. After 24 hours examine the membrane for typical colonies which are 2–3 mm in diameter and pale orange in colour. Colonies are smooth, entire and circular. Other coliforms produce pink colonies from the fermentation of sorbitol.

Note: Recovery of *E. coli* 0157:H7 from enrichment broth is improved by using immunomagnetic separation with antibody-coated magnetic beads, Wright, Chapman and Siddons (1994). Non-specific binding of other organisms is reduced by washing beads with phosphate buffered saline containing 0.002–0.05 % tween 80.

8.7.4.4 Confirmation

Subculture sorbitol-negative colonies on suitable non-selective NA and incubate at 37°C for 24 hours. Examine plates for purity and if cultures are not pure, subculture a selection of colonies on fresh NA for a further 24 hours at 37°C. Check that isolates are oxidase negative.

Isolates may be confirmed by slide agglutination using a specific commercially prepared latex suspension. Separately, place two drops of saline or phosphate-buffered saline on a microscope slide. Emulsify the same colony of the suspected culture in each drop to give a smooth, fairly dense suspension. Check to ensure that the suspension is smooth and that auto-agglutination has not occurred. Mix one loopful of latex reagent to one suspension and rock the slide gently. Examine for evidence of agglutination. Repeat the test with the appropriate positive and negative controls. Isolates showing strong positive agglutination when compared with the positive control should be considered positive.

Isolates should be confirmed as *E. coli* by appropriate biochemical tests (Cowan 1993) or use of a suitable test kit.

8.8 *Salmonella*

8.8.1 Introduction

Many different serotypes of salmonellas are present in wild and domestic animals and agricultural livestock and any of these sources may act as reservoirs of human salmonellosis. Salmonellas in the environment are derived from acute and convalescent cases, from chronic carriers and symptomless faecal excreters in human and animal populations.

Industrial effluents from abattoirs and animal processing factories contribute salmonellas to sewage and thence to the local environment and water courses.

8.8.2 Principle

Concentration of organisms on a membrane filter or by means of a filter-aid followed by culture in a non-selective pre-enrichment medium. Subculture to a selective enrichment broth followed by plating on selective agar and biochemical and serological confirmation of characteristic colonies.

8.8.3 Definition

Salmonellas normally conform to the general definition of the Family Enterobacteriaceae (Le Minor, Rohde and Taylor 1970) and can be further differentiated, biochemically, into four groups, subgenus I to IV. Those of subgenus I, the largest group and the only ones considered to be pathogenic for humans are ß-galactosidase-negative.

The usual biochemical reactions include production of gas from glucose, and hydrogen sulphide production. Indole and urease are not produced. Citrate is utilised as a sole carbon source; lysine and ornithine are decarboxylated. Phenylalanine and tryptophan are not oxidatively deaminated; sucrose, salicin, inositol and amygdalin are not fermented.

Salmonellas are subdivided into serovars on the basis of genus-specific combinations of somatic and flagellar antigens. They may be further differentiated by bacteriophage and plasmid typing.

8.8.4 Choice of media

Enrichment media either suppress the growth of other enterobacteria, allowing salmonellas to thrive, or they enhance the growth of salmonellas, allowing them to predominate in a mixed culture. Media selective for *Salmonella* species contain inhibitory substances such as selenium salts, brilliant green or malachite green. After incubation, the selective enrichment cultures are plated on differential agar media on which salmonellas form colonies with a characteristic appearance and morphology. Representative colonies are subcultured for biochemical and serological testing.

The following media may be used in these methods:

(i) Buffered Peptone Water (BPW) (Edel and Kampelmacher 1973) for pre-enrichment.

(ii) Selenite Broth (SB) (Hobbs and Allison 1945a,1945b) for the selective enrichment of *Salmonella typhi* and *S. paratyphi B*.

(iii) Rappaport Vassiliadis Broth (RVB) (Vassiliadis et al 1976, Vassiliadis 1983) for the selective enrichment of other salmonellas. This medium and a modification containing soy peptone instead of tryptone and a lower concentration of magnesium chloride are commercially available as dehydrated products.

(iv) Xylose Lysine Desoxycholate Agar (XLDA) (Taylor and Harris 1965) for selective isolation of salmonellas.

(v) Bismuth Sulphite Agar (BSA), modified from Wilson and Blair (1927).

(vi) Brilliant Green Agar (BGA) (ICMSF 1978). A modified version of this medium is commercially available as a dehydrated product. BGA is not recommended for the isolation of *S. typhi*.

(vii) Nutrient Agar Slopes.

(viii) Lysine Iron Agar slopes (LIAS) (Edwards and Fife 1961) or Triple Sugar Iron Agar slopes (TSIAS) and Urea Broth (UB), for preliminary biochemical identification.

8.8.5 Safety

Selenium salts are toxic if inhaled or ingested and there is a possible risk of teratogenicity in pregnant laboratory workers.

In laboratories of a containment level inappropriate for the chance isolation of *Salmonella typhi*, (see section 6.1 and especially sections 6.1.2 and 6.1.3) media containing selenite should not be included in the range of selective enrichment media.

8.8.6 Practical considerations

Undue delay between the time the sample is taken and processed in the laboratory will reduce the chances of isolating salmonellas.

Selective media may be too inhibitory for organisms stressed in the environment and pre-enrichment in a non-inhibitory medium is necessary before transfer to liquid selective enrichment media.

Salmonella isolation may be improved if several small sub-samples are examined rather than one large portion.

Dilution of the material reduces the effect of substances toxic to salmonellas.

Competition from other organisms, particularly the *Pseudomonas* group, may reduce isolation efficiency. The technique of incorporating a final concentration of 5 mgL^{-1} of cefsulodin in the pre-enrichment stage will eliminate the *Pseudomonas* species but *Serratia* species may flourish (Harmon, Stuart and Mawer 1991).

In any microbiological method, quality control is important to ensure that the culture media, both liquid and solid, are able to support the growth of low numbers of the micro-organisms under investigation. The use of old stock culture solutions of *Salmonella* is not recommended for the quality control of salmonella plating media. Old stock solutions tend not to produce typical colonial appearances.

8.8.7 Pre-enrichment procedure

8.8.7.1 Membrane filtration

Filter the sample as described in section 7.5, using membranes of pore size 0.45 µm. Place each membrane in approximately 100 mL of enrichment medium in a wide-mouthed screw-capped container.

8.8.7.2 Filtration with filter-aid

8.8.7.2.1 For qualitative examination Filter the sample as described in sections 7.5 and 8.4 and transfer the absorbent pad with filter-aid to approximately 100 mL of enrichment medium in a wide-mouthed container. Any filter-aid adhering to the funnel is rinsed into the container with the medium.

8.8.7.2.2 For quantitative examination Filter a measured volume of sample as described in section 8.4. Place the filter pad with filter-aid in 100 mL of enrichment medium. Rinse all the filter-aid into the medium. Mix well and transfer 5 separate volumes of 10 mL to sterile tubes, keeping the remaining 50 mL volume as a separate culture. If high counts are expected, pipette 5 × 1 mL and 5 × 0.1 mL volumes from the 50 mL volume to separate tubes each containing 10 mL of enrichment medium.

8.8.8 Enrichment procedure

8.8.8.1 Isolation of salmonella species other than S. typhi and S. paratyphi

8.8.8.1.1 Incubation of buffered peptone water. Incubate the containers of BPW with membranes or filter-aid for a resuscitation period of 24 ± 2 hours at 37°C.

8.8.8.1.2 For qualitative examination. Subculture 0.1 mL from 100 mL BPW pre-enrichment culture to 10 mL of RVB. Incubate RVB at 41°C for up to 48 hours.

8.8.8.1.3 For quantitative examination. Subculture 0.1 mL from each BPW pre-enrichment culture to 10 mL of RVB. Incubate RVB as above.

8.8.8.2 Isolation of S. typhi and S. paratyphi B

Incubate the containers of SB with membranes or filter-aid at 37°C for up to 48 hours.

8.8.9 Subculture of enrichment media to selective agars

After incubation for 24 and 48 hours subculture the RVB or SB to XLDA and to either BGA or BSA. BGA should not be used for the isolation of *S. typhi*.

Incubate the selective agars at 37°C for 18 to 24 hours.

8.8.10 Isolation and identification on selective agar

Examine the plates for characteristic colonies.

8.8.10.1 Xylose Lysine Desoxycholate Agar

Salmonellas produce red colonies, usually with a black centre. Xylose-fermenting coliforms produce yellow colonies. *Pseudomonas* species produce either red colonies or yellow colonies with grey-black centres. Shigellas produce small pink-red colonies. *Proteus* produces irregular pink colonies with small black centres.

Salmonella colonies may be obscured by overgrowth and this medium should be used in conjunction with BGA or BSA.

8.8.10.2 Brilliant Green Agar

Salmonellas produce red colonies and lactose- and sucrose-fermenting coliforms produce yellow/green colonies. The growth of *Proteus* species is usually inhibited, but when it does occur, colonies are red. Similarly the growth of *Pseudomonas* species is usually inhibited but small crenated colonies may be produced.

8.8.10.3 Bismuth Sulphite Agar

Salmonellas produce black-centred colonies with a clear periphery. The ratio of black centre to clear edge varies and the colony may be surrounded by a metallic sheen (McCoy and Spain 1969). Some strains of *Salmonella* produce a green colony with a darker green centre. Coliforms produce large brown colonies with no clear periphery. *Proteus* species produce irregular green-brown colonies and *Pseudomonas* species produce small green colonies with a sticky consistency.

8.8.11 Confirmation

8.8.11.1 Biochemical confirmation

Subculture characteristic colonies from each plate, using a straight wire, to confirmatory media such as LIAS or TSIAS, together with UB and NA as a check for purity. The straight wire carrying the organisms from the colony should be stabbed into the butt and streaked along the slant as it is withdrawn. Avoid stabbing through the butt to the bottom of the tube; the end of the wire must remain approximately 3 mm from the bottom of the tube as gas production may cause the medium to be blown out of the tube. Incubate at 37°C for 18–24 hours. Regard cultures which give characteristic reactions in these confirmatory media as presumptive salmonellas (see Tables 5 and 6). Alternatively one of the commercially available identification systems may be used.

Table 5 Reactions in Lysine Iron Agar

Genus	Slope	Butt	H_2S
Arizona	Alk	Alk	+
Salmonella	Alk	Alk	+
Proteus	Red	A	+ or –
Providencia	Red	A	–
Citrobacter	Alk	A	+
Escherichia	Alk	A or NC	–
Shigella	Alk	A	–
Klebsiella	Alk	Alk	–
Enterobacter	Alk	A	–

A = Acid (yellow)
Alk = Alkaline (purple)
NC = No change
H_2S+ = Blackening

Table 6 Reactions in Triple Sugar Iron Agar and Urea Broth

Genus	Slope	Butt	H_2S	Urea Broth
Klebsiella	A	AG	–	+ or –
Enterobacter	A	AG	–	–
Escherichia	A	AG	–	–
Proteus vulgaris	A	AG	+	+
Proteus morganii	NC or Alk	A or AG	–	+
Shigella	NC or Alk	A	–	–
S. typhi	NC or Alk	A	+ (weak)	–
Salmonella (other)	NC or Alk	AG	+	–

A = Acid (yellow)
AG = Acid (yellow) and gas formation
NC = No change
Alk = Alkaline (red)
H_2S + = Blackening – =No change
Urea Broth + = Alkaline (red) – =No change

8.8.11.2 Serological confirmation

Subculture characteristic colonies on to moist NA slopes.

For optimum flagella formation, it is essential that fluid is present in the tube and sterile broth should be added if required. Incubate overnight at 37°C. Carry out a slide agglutination as follows:

Place 3 separate drops (0.02 mL) of saline onto a clean microscope slide, using a wire loop or pipette. Emulsify growth from the moist butt of the slope in each separate drop to produce homogeneous suspensions.

Mix a loopful of *salmonella* Polyvalent 'O' (PSO) antiserum with the first drop of suspension and a loopful of *salmonella* Polyvalent 'H' (PSH) serum with the second. Gently rock the slide back and forth and examine for agglutination against a black background under a bench lamp. The third drop containing no antiserum indicates whether or not the culture auto-agglutinates. These strains should be re-plated on XLDA, BGA or BSA and dry smooth colonies dealt with as previously described.

Organisms which agglutinate with PSO and PSH antisera or strains which agglutinate with PSH serum only can be presumptively regarded as members of the *Salmonella* group. It is advisable to check colonially characteristic PSO-negative, PSH-positive or PSH-negative strains for agglutination with Salmonella Vi antiserum to exclude *Salmonella typhi* as the Vi antigen may mask the 'O' and 'H' antigens.

Cultures should be submitted to a reference centre for final confirmation and serotyping.

8.9 *Shigella*

8.9.1 Introduction

Members of the genus *Shigella* are intestinal pathogens of humans, causing dysentery or shigellosis, and may also be carried asymptomatically in the intestinal tract.

There are four species, *S. dysenteriae*, *S. sonnei*, *S. flexneri* and *S. boydii* all of which cause gastrointestinal disease. The commonest species isolated in the United Kingdom is *S. sonnei*, although the most severe disease is caused by *S. dysenteriae*, type 1 which produces a potent exotoxin (Shiga toxin).

8.9.2 Principle

Isolation of organisms on a membrane followed by enrichment in a selective medium and plating onto a selective agar. Characteristic colonies are confirmed by slide agglutination and may be sub-cultured onto a non-selective agar for further biochemical tests or serotyping.

8.9.3 Definition

Bacteria in the genus *Shigella* are Gram-negative non-motile rods which are oxidase-negative and catalase-positive (with the exception of *S. dysenteriae* type 1, which is catalase-negative). Citrate cannot be used as a sole carbon source, and sugars are fermented without gas production. (A few strains however produce gas).

8.9.4 Choice of media

The following media are recommended for isolation of shigella from water:

(i) A modification of Hajna GN Enrichment Broth (HEB).

(ii) A modification of Desoxycholate Citrate Agar (MDCA).

(iii) Hektoen Agar (HA) (King and Metzger 1968).

8.9.5 Procedure

8.9.5.1 Membrane filtration

Filter the sample as described in section 7.5, using a membrane of pore size 0.45 µm. It may be necessary to use several membranes if the water is turbid.

8.9.5.2 Enrichment

Place the membrane(s) in 100 mL of HEB in a suitable screw-capped container. Shake the broth and incubate at 37°C for 6–8 hours.

8.9.5.3 Isolation and identification on selective agar

Shake the enrichment broth before subculturing to HA and MDCA. Incubate the agar plates at 37°C for 18–24 hours. Characteristic non-lactose-fermenting colonies are subcultured on a non-selective agar such as NA and incubated at 37°C for 18–24 hours.

Note: Those working on the isolation of shigellas should familiarise themselves with the type of growth and colonial morphology of these organisms on HA and MDCA agars (see Tables 7 and 8). Plating on selective agar must produce isolated colonies. Failure to do so may result in 'false negatives' since typical colonies may be overgrown. Where this occurs or colonial identification is uncertain, a subculture to fresh agar should be performed.

Table 7 Colonial appearance on HA after incubation at 37°C for 18–24 hours

Organism	Characteristic Appearance
Shigella	Moist green colonies, 2–4 mm diameter. (*Shigella sonnei* may produce larger irregularly shaped colonies).
Salmonella	Blue-green colonies, with or without black centres.
Pseudomonas	Large rough textured green colonies.
Enterobacter, Citrobacter, Escherichia, Klebsiella	Yellow colonies. The medium around the colonies often turns salmon pink.
Proteus	Pale green or ochre yellow colonies.

Table 8 Colonial appearance on MDCA after incubation at 37°C for 18–24 hours

Organism	Characteristic Appearance
Shigella	Small raised cream coloured colonies.
Salmonella	Large black colonies with a thin white periphery.
Pseudomonas	Very small flat cream coloured colonies.
Proteus	Cream colonies with small black centre.
Coliforms	Pink mucoid colonies with raised centres.
Escherichia	Pale pink colonies with grey centres.

8.9.6 Confirmation

8.9.6.1 Biochemical confirmation

Non-lactose-fermenting colonies which have been subcultured onto a non-selective agar may be further identified using one of the commercially available multi-test biochemical identification kits.

8.9.6.2 Serological confirmation

Colonies which identify biochemically as shigellas must be confirmed serologically, using polyvalent anti sera and single-factor sera.

Place two separate drops (0.02 mL) of saline on a glass slide. Emulsify suspected culture with a loop into each drop of saline to give a smooth, fairly dense suspension. Add one drop of undiluted serum to one suspension and mix. Rock the slide gently and observe for agglutination using indirect lighting over a dark background. Agglutination should be strong and clearly visible within one minute. Non-specific agglutination due to minor antigenic reactions may occur and these sometimes are usually slow to appear. This will however be apparent in the second suspension, which is used as a control.

8.10 Staphylococci

8.10.1 Introduction

Staphylococci, particularly *Staphylococcus aureus*, are of concern in the hospital environment because of their potential to cause infections. They have also been advocated as indicators of the quality of bathing water including swimming pools, hydrotherapy pools and sea water. They are not normally present in drinking water supplies but their detection may be required in food and pharmaceutical manufacture and in hospitals.

8.10.2 Definition

Members of the genus *Staphylococcus* are Gram-positive cocci, facultatively anaerobic, non-motile, do not form spores and are usually catalase-positive and sensitive to lysostaphin. In the context of the method, staphylococci are defined as fermenting mannitol and able to grow in the presence of 7.5% sodium chloride and 0.005% sodium azide.

8.10.3 Principle

Filtration of the test volume of sample and incubation on a selective agar medium containing sodium azide, sodium chloride, mannitol and phenol red. Presumptive isolates should be subcultured to confirm identity.

8.10.4 Choice of medium

The method described utilises the medium M-5LSMA developed by Stengren and Starzyk (1984). This medium is lipovitellin-salt-mannitol agar modified by the inclusion of 0.005% sodium azide (see section 7.9.3) to suppress the growth of Gram-positive bacilli. The method is primarily directed at detecting *Staph aureus*. A few species, for example *Staph epidermidis* and *Staph hominis* do not ferment mannitol and are not readily detected. The medium also contains egg yolk to enable lipase activity to be detected by the production of an opaque zone around the colony. This is difficult to see under a membrane and can therefore be omitted when the medium is used for membrane filtration alone.

8.10.5 Procedure

After preparing the sample and making any required dilutions as described in section 7.4, filter the test volume as in section 7.5 and place each membrane on the agar surface in a Petri dish containing M-5LSMA. Incubate the plates aerobically at 29–30°C for 48 hours. Count all colonies that are yellow, opaque off-white or pale yellow, and record as presumptive staphylococci.

8.10.6 Confirmation

To confirm strains as *Staphylococcus* species, subculture each colony to be tested (see 6.3.3) to NA. After incubation for 18–24 hours at 30–37°C check growth for purity and carry out a Gram-stain. Further identification to species level may be performed using an appropriate commercial identification kit. To confirm isolates as *Staph aureus* it is necessary to test either for coagulase production or to re-streak on M-5LSMA. Opaque zones around colonies after 24 hours indicate lipovitellin-lipase activity which correlates well with coagulase production. Yellow zones around the colonies after 24–48 hours indicate mannitol fermentation. *Staph aureus* ferments mannitol and also produces lipovitellin-lipase and coagulase.

8.11 *Vibrio cholerae* and Other *Vibrio* Species

8.11.1 Introduction

The species *Vibrio cholerae* can be subdivided into over 80 O-serovars. The organisms that usually cause epidemic cholera are toxin producing strains of the O1 Serovar; however, recently a new Serovar, 0139, has been identified as having epidemic potential. Strains of other serovars (non-O1 *V.cholerae*) can also cause diarrhoea either by means of cholera toxin or, more commonly, by other enteropathogenic mechanisms. Other species of *Vibrio*, in particular *V. parahaemolyticus*, *V. fluvialis* and *V. mimicus*, can also cause diarrhoea. All *Vibrio* species are natural inhabitants of brackish, estuarine or coastal water. All require salt for optimum growth and survival, and the degree of requirement reflects the ionic conditions of their natural ecological niche; *V. cholerae* has the least requirement and *V. parahaemolyticus*, being of coastal origin, has the greatest. Potable water can be a vehicle for *V. cholerae* and carriage of other *Vibrio* species is possible. No species of *Vibrio* should be present in potable water.

8.11.2 Definition

Members of the Genus *Vibrio* species are Gram negative curved or comma-shaped rods, actively motile, aerobic, facultatively anaerobic, oxidase-positive, able to grow at pH 8.6 and inhibited by 0/129 phosphate.

8.11.3 Principle

Vibrio cholerae and most other pathogenic *Vibrio* species are tolerant of alkaline conditions and are able to grow rapidly at 37°C in simple peptone water containing sufficient sodium chloride to encourage growth. These properties are utilised for the enrichment of the organisms in alkaline peptone water (1% peptone, 1% sodium chloride and pH 8.6). A concentration of 1% sodium chloride gives good growth of all species likely to be encountered, although 0.5% sodium chloride is adequate for *V. cholerae*.

After overnight enrichment at 37°C, broths are subcultured to the selective medium, thiosulphate-citrate-bile salt-sucrose agar (TCBSA) (Kobayashi et al 1963). This medium was developed for the isolation of *V. parahaemolyticus* but has proved useful for the isolation of most other vibrios, especially *V. cholerae*. It has achieved wide acceptance as the medium of choice for the isolation of vibrios because it is highly selective, readily available commercially, easy to prepare (requiring no autoclaving) and permits the preliminary differentiation of the sucrose-fermenting (yellow colonies) species, such as *V. cholerae* and *V. fluvialis,* from the non-sucrose-fermenting species (blue green colonies), such as *V. parahaemolyticus* and *V. mimicus*.

8.11.4 Procedure

Add at least 100 mL of sample to an equal volume of double-strength alkaline peptone water. Incubate at 25°C for 2 hours followed by 12–16 hours at 37°C.

After incubation, subculture from the top of the enrichment broth to a plate of TCBSA and incubate the plate at 37°C for 16–24 hours.

Examine the plates for vibrio-like colonies and subculture at least three of each colony type to NA for a purity check, oxidase test and further identification. Colonies of vibrios will usually be of 1–3 mm diameter and either yellow (sucrose-fermenting) or blue-green (non-sucrose-fermenting).

Strains that are oxidase-positive require further identification. If epidemic strains of *V. cholerae* 01 are sought, then colonies that are both sucrose-fermenting (yellow on TCBSA) and oxidase-positive can be checked for agglutination with *V. cholerae* O1 antiserum. A strain that agglutinates can be identified as a presumptive isolate of *V. cholerae* O1 and should be sent to the appropriate reference laboratory for confirmation.

8.11.5 Enumeration

If enumeration is required, the multiple tube method can be used. Inoculate 1 × 50 mL and 5 × 10 mL amounts of sample into equal volumes of double-strength alkaline peptone water and proceed as above. Volumes of 1 mL or less should be added to 20 mL volumes of single-strength alkaline peptone water.

8.11.6 Membrane filtration

Direct enumeration by filtration and incubation of the filter on the surface of a TCBSA plate is not satisfactory due to the overgrowth of the membranes by other bacteria. Filtration can be used as a means of concentration by placing the filter in single strength alkaline peptone water for enrichment and proceeding as above.

8.11.7 Confirmation

Commercial identification kits can be used to identify oxidase-positive colonies.

8.12 *Yersinia*

8.12.1 Introduction

The isolation procedure described gives good recovery of *Yersinia* species from raw waters.

8.12.2 Definition

Yersinias are Gram-negative rods, non-motile at 37°C (but some species are motile at 22°C), catalase-positive, oxidase-negative and facultatively anaerobic. Sugars are attacked fermentatively with occasional gas production. Virulent strains are aesculin-negative and can bind crystal violet dye.

In the context of the method, organisms are regarded as yersinias which, after enrichment and alkali treatment, give characteristic colonies on Cefsulodin Irgasan Novobiocin agar (CINA) (Mair and Fox 1986, based on Schiemann 1982), and which then give positive reactions with TSIAS and Urea Broth, and are non-motile at 37°C after 4–6 hours. Further biochemical and serological identification should be carried out to speciate the isolates.

8.12.3 Principle

Concentration of organisms on a membrane filter followed by enrichment. Incubation and alkali treatment of enrichment broth, followed by plating on CINA. Biochemical and serological confirmation of characteristic colonies.

8.12.4 Choice of media for isolation

Yersinia organisms are poor competitors, and the method employed for enrichment may require adjustment depending upon the degree of contamination of the sample, and the timescale regarded as acceptable for isolation.

The enrichment procedure uses Tris-buffered peptone which is incubated at 9°C for up to two weeks. The first choice of a selective agar is CINA which is available as a commercially prepared dehydrated product.

8.12.5 Identification media

The following media may be used:

(i) Triple Sugar Iron agar slopes (TSIAS);

(ii) Urea Broth (UB);

(iii) MacConkey Agar (MA).

8.12.6 Procedure (Greenwood 1993)

8.12.6.1 Enrichment

After filtration of the sample as described in section 7.5, place membrane(s) in 100 mL of tris-buffered peptone in screw-capped containers and incubate at 9°C for two weeks.

8.12.6.2 Alkali treatment

Prepare 0.5% m/V potassium hydroxide in 0.5% saline. To 9 volumes of this, add 1 volume of the incubated enrichment broth, subculture to CINA and incubate at 30°C for 24 hours.

8.12.6.3 Isolation on selective agar

Subculture typical 'bullseye' colonies isolated on CINA to each of the following media and incubate at 30°C for the specified times:

(i) TSIAS—24 hours;

(ii) UB—1 to 3 days;

(iii) MA—24 hours;

Cultures from the MA plate which are urease-positive and produce acid but no gas and which do not produce H_2S on TSIAS are subcultured from the MA plate to NB.

8.12.6.4 Motility test

Incubate the nutrient broth cultures at 37°C for 4–6 hours. Cultures which do not exhibit motility are assumed to be *Yersinia* species. Yersinias are motile at temperatures below 28°C.

8.12.6.5 Speciation

Speciation can be carried out using commercially available identification kits.

APPENDICES

Appendix A: Laboratory Glassware and Plastic ware:

Specifications:

Sample bottles

See section 6.4.1.

Pipettes

Glass pipettes should conform to BS (BSI 1976:700). The 1 mL and 10 mL sizes should be the straight-sided, total delivery type. Alternatively, appropriately calibrated automatic pipettes with sterile disposable tips may be used. Plastic disposable pipettes may also be useful.

Glassware for liquid media

The 50 mL volumes of double-strength medium required for the multiple tube method should be distributed in screw-capped bottles of at least 125 mL capacity, and the 10 mL volumes in test-tubes with a capacity of about 30 mL. For the 5 mL volumes of single-strength medium and for most other liquid media, tubes with a capacity of about 22 mL or screw-capped bottles are suitable. When inner (Durham) fermentation tubes are required, suitable sizes for the 5 and 10 mL volumes, and for the 50 mL volumes of medium are 35 × 8 mm and 75 × 10 mm respectively. When media have to be stored for several weeks, screw-capped containers should be used.

Petri dishes

It is important that Petri dishes used for colony counts should be of a standard size to ensure that the surface area and depth of the medium are always constant.

Cleaning and Preparation

Where applicable, after use, pipettes should be placed in jars of disinfectant, for example hypochlorite solution containing 0.1% of free chlorine, left for one hour and rinsed several times. After a final rinse in distilled water, pipettes should be dried and packed in metal canisters for sterilization. Plugging pipettes with non-absorbent cotton wool may help to reduce cross-contamination. It is advisable to clean glass pipettes periodically with a non-toxic detergent or acid and then rinse thoroughly.

Test tubes, bottles and flasks should be autoclaved after use and cleaned with a brush, washed in water with a non-toxic detergent, rinsed in clean water and finally in distilled water.

Test tubes and flasks should be plugged with non-absorbent cotton wool or covered with closely-fitting aluminium or polypropylene caps. The necks and stoppers of bottles should be covered with paper or foil to prevent contamination.

Sterilization of Glassware

Glassware may be sterilized in hot-air ovens at 160°C for 1 hour. If this temperature is exceeded, the exposure time may be reduced proportionately. A temperature above 170°C should be avoided since organic matter tends to char and cotton wool is rendered friable. Alternatively, bottles and tubes may be autoclaved at 121°C for 20 minutes. If bottles with ground-glass stoppers are used, a strip of paper or foil approximately 75 × 10 mm should be inserted between the stopper and the neck of the bottle before sterilization. This prevents jamming of the stopper and cracking of the glass on cooling.

Appendix B: Media and Reagents:

Choice of Constituents:

Peptone

In the preparation of media, unless a particular grade is specified, any microbiological peptone may be used which forms a clear solution and does not precipitate when alkali is added to adjust the pH.

Agar

In the instructions which follow, a satisfactory gel strength with the concentration of agar in the medium is assumed. This may vary according to the quality of the agar available. It is also assumed that the agar forms a solution which remains sufficiently clear for filtration to be unnecessary.

Bile salts

The term bile salts includes sodium taurocholate and sodium tauroglycocholate. Different preparations of bile salts vary in their inhibitory properties; each new batch should be tested against a known satisfactory product and the concentration adjusted accordingly.

Distilled water

Where distilled water is specified, a glass still should be used; alternatively, deionised water may also be used.

Dehydrated media

Some of the media described are available commercially in dehydrated form and should be reconstituted according to the manufacturers' instructions.

Sterilization of media

A time-temperature combination of 121°C for 15 minutes is usually specified for many microbiological purposes. However, a temperature of 115°C for a minimum of 10 minutes is recommended for most media used in water examination. The heating profile of the autoclave should be established for typical loads. This information may then be used to specify the sterilization regime for different volumes of media.

Storage of media

In general, after sterilization most media in sealed containers may be stored safely for several months at room temperature provided they are kept in the dark. Media dispensed aseptically may be kept at 4–10°C for up to one month; before use, media should be inspected carefully for any untoward signs such as contamination or excessive evaporation. Most reagents are best kept at 4°C.

Preparation of Media and Reagents

pH adjustment

The pH of media is of vital importance. Measurement and, where necessary, adjustment of pH is part of the preparation of most media. Where a method of adjustment is not specified, general guidance can be found in 'Collins and Lyne's Microbiological Methods' (Collins 1989).

BASIC MEDIA

Nutrient Broth

Meat extract	10 g
Peptone	10 g
Sodium chloride	5 g
Distilled water	1 litre

Add the ingredients to the water and heat to dissolve. Adjust the pH to about 8.2 with a solution of molar sodium hydroxide and boil for 10 minutes. Clarify by filtration and adjust pH to between 7.2–7.4. Dispense in bottles or tubes and autoclave at 115°C for 10 minutes.

Nutrient Agar

Nutrient Broth gelled by the addition of agar, usually in the order of 1–2 % m/V.

Blood Agar

Nutrient Agar with the addition of 5% (V/V) sterile horse blood.

MEDIA FOR COLIFORM ORGANISMS

Membrane Lauryl Sulphate Broth (PHLS/SCA 1980a)

Peptone	40 g
Yeast Extract	6 g
Lactose	30 g
Phenol red (0.4% m/V aqueous solution)	50 mL
Sodium lauryl sulphate—specially pure	1 g
Distilled water	1 litre

Add the ingredients to the water and mix gently to avoid froth. The final pH of the medium should be 7.4–7.5 and it may be necessary to adjust the pH to about 7.6 before sterilization to achieve this. Distribute in screw-capped bottles and autoclave at 115°C for 10 minutes. (The media may need to be removed from the autoclave as soon as possible to avoid breakdown of the lactose and a reduced pH).

Improved Formate Lactose Glutamate Medium (PHLS 1969, modified from Gray 1964)

This is commercially available in dehydrated form as Minerals Modified Glutamate Medium, which may require pH adjustment.
Double-strength medium:

Lactose	20.0 g
L(+) Glutamic acid sodium salt	12.7 g
L(+) Arginine monohydrochloride	40 mg
L(−) Aspartic acid	48 mg
L(−) Cystine	40 mg
Sodium formate	500 mg
Dipotassium hydrogen phosphate	1.8 g
Ammonium chloride	5.0 g
Magnesium sulphate heptahydrate	200 mg
Calcium chloride dihydrate	20 mg
Iron(III) citrate scales	20 mg
Thiamine (Aneurin hydrochloride)	2 mg
Nicotinic acid	2 mg
Pantothenic acid	2 mg
Bromocresol purple (1% m/V ethanolic solution)	2 mL
Distilled water to	1 litre

The medium is most conveniently prepared in quantities of 10 litres or more. If it is not to be distributed in tubes immediately, the lactose and thiamine should be omitted and added before dispensing. Several of the ingredients are more conveniently added as separate solutions and these may be prepared as follows:

SOLUTION 1	
L(+) Arginine monohydrochloride	400 mg
L(−) Aspartic acid	480 mg
Distilled water	50 mL
Heat to dissolve.	

SOLUTION 2	
L(−) Cystine	400 mg
5M Sodium hydroxide	10 mL
Distilled water	90 mL
Heat to dissolve.	

SOLUTION 3	
Nicotinic acid	20 mg
Pantothenic acid	20 mg
Distilled water	5 mL
Dissolve in the cold.	

SOLUTION 4	
Iron(III) citrate scales	200 mg
Distilled water	10 mL
Heat to dissolve.	

SOLUTION 5	
Calcium chloride dihydrate	5 g
Distilled water	100 mL
Concentrated hydrochloric acid	0.1 mL

Dissolve in the cold and sterilize at 121°C for 20 minutes.
Keep as a stock solution.

SOLUTION 6
Sterile 0.1% solution of thiamine in distilled water. Add the contents of an ampoule (100 mg) to 99 mL of sterile distilled water. The solution should be kept at 4°C and any remaining discarded after 6 weeks.

To prepare 10 litres of double-strength medium, dissolve the appropriate quantities of L(+) glutamic acid sodium salt, sodium formate, dipotassium hydrogen phosphate, ammonium chloride and magnesium sulphate in 9 litres of hot distilled water. Add the whole of solutions 1,2,3 and 4, and 4 mL of solution 5. Adjust the pH to 6.8 or higher if necessary, so that the final pH, after completion and sterilization, is 6.7.

After adjustment of pH, add 20 mL of 1% ethanolic solution of bromocresol purple. Make up the final volume to 10 litres. This should require about another 810 mL of distilled water. If the bulk of the medium is not required for immediate use, store in 500 mL volumes and autoclave at 115°C for 10 minutes. For use, add the necessary amount of lactose and thiamine (solution 6), allow to dissolve and distribute in 10 mL and 50 mL volumes. Each tube or bottle should contain an inverted (Durham) fermentation tube. Sterilize at 115°C for 10 minutes or steam at 100°C for 30 minutes on three successive days.

Single-strength medium:
Prepare single-strength medium by diluting the double-strength medium with an equal volume of distilled water and distribute in 5 mL volumes in tubes containing an inverted fermentation (Durham) tube. Sterilize at 115°C for 10 minutes or steam at 100°C for 30 minutes on three successive days.

Lauryl Tryptose Lactose Broth (APHA 1989)

Double-strength medium:

Tryptose	40.0 g
Lactose	10.0 g
Sodium chloride	10.0 g
Dipotassium hydrogen phosphate	5.5 g
Potassium dihydrogen phosphate	5.5 g
Sodium lauryl sulphate—specially pure	200 mg
Distilled water	1 litre

Add the tryptose, sodium chloride, lactose and phosphates to the water and warm to dissolve. Add the sodium lauryl sulphate and mix gently to avoid froth. Adjust to pH 6.8. Prepare single-strength medium by dilution of the double-strength medium with an equal volume of distilled water.

Distribute single-strength medium in 5 mL volumes and double-strength medium in 10 mL and 50 mL volumes. Each tube or bottle should contain an inverted fermentation (Durham) tube. Autoclave at 115°C for 10 minutes.

Lauryl Tryptose Mannitol Broth with Tryptophan (PHLS/SCA 1980c)

Tryptose	20 g
D(–) Mannitol	5 g
Sodium chloride	5 g
Dipotassium hydrogen phosphate	2750 mg
Potassium dihydrogen phosphate	2750 mg
Sodium lauryl sulphate	100 mg
L(–) Tryptophan	200 mg
Distilled water	1 litre

Add the tryptose, sodium chloride, mannitol, phosphates and tryptophan to the water and warm to dissolve. Add the sodium lauryl sulphate and mix gently to avoid froth. Adjust to pH 6.8. Distribute in 5 mL volumes in tubes containing an inverted fermentation (Durham) tube. Autoclave at 115°C for 10 minutes.

Brilliant Green Lactose Bile Broth

Peptone	10 g
Lactose	10 g
Ox bile (dehydrated)	20 g
Brilliant green (0.1% m/V aqueous solution)	13 mL
Distilled water to	1 litre

Dissolve the peptone in 500 mL distilled water. Add the dehydrated ox bile dissolved in 200 mL of distilled water; this solution should have a pH between 7.0–7.5. Make up with distilled water to approximately 975 mL. Add the lactose and adjust the pH to 7.4. Add the brilliant green solution and make up the volume with distilled water to 1 litre.

Distribute 5 mL volumes in test tubes containing inverted fermentation (Durham) tubes and autoclave at 115°C for 10 minutes.

Tryptone Water for Indole Reaction

Certain peptones which give satisfactory results in tests at 37°C are not satisfactory for the indole test at 44°C (Burman 1955). Care should be taken in the appropriate selection of reagents.

Tryptone	20 g
Sodium chloride	5 g
Distilled water	1 litre

Dissolve the ingredients in the water and adjust to pH 7.5. Distribute in 5 mL volumes and autoclave at 115°C for 10 minutes.

MacConkey Agar

Bile salts	5 g
Peptone	20 g
Lactose	10 g
Sodium chloride	5 g
Agar	12 g
Neutral red (1% m/V aqueous solution)	5 mL
Distilled water	1 litre

Add the ingredients to the water and steam to dissolve. Distribute into screw-capped bottles and autoclave at 115°C for 10 minutes. For use, melt in steam and pour into sterile Petri dishes, using 15 mL of medium in each dish.

Lactose Peptone Water

Peptone	10 g
Sodium chloride	5 g
Lactose	10 g
Phenol red (0.4% m/V aqueous solution)	2.5 mL
(or Andrade's indicator)	(10 mL)
Distilled water	1 litre

Dissolve the ingredients in the water and adjust to pH 7.5. Add the phenol red indicator and distribute in 5 mL volumes into tubes containing inverted fermentation (Durham) tubes. Alternatively, adjust pH to between 6.8–7.0 and add the Andrade's indicator. Autoclave at 110°C for 10 minutes. Alternatively, steam (100°C) for 20 minutes on each of three successive days. Test for sterility by incubation at 37°C for 24 hours.

Andrade's Indicator

This is prepared by dissolving 0.5 g of acid fuchsin in 100 mL of distilled water. Add 17 mL of 1M sodium hydroxide solution and leave at room temperature overnight when the solution should appear straw coloured. If any brownish colour appears, add a little more sodium hydroxide solution and allow to stand again. This solution is strongly alkaline, and consequently, media to which it is added should be adjusted previously to a pH of about 6.8.

MEDIA FOR FAECAL STREPTOCOCCI

NOTE Sodium azide is highly toxic if ingested or inhaled and care must be taken when handling it. Solutions containing azide should not be discharged through metal pipework or drains as explosive compounds may be formed. Azides can be decomposed by treatment with an excess of nitrite solution.

Membrane Enterococcus Agar (Slanetz and Bartley 1957)

Tryptose	20 g
Yeast extract	5 g
Glucose	2 g
Dipotassium hydrogen phosphate	4 g
Sodium azide	400 mg
Agar	12 g
2,3,5-triphenyltetrazolium chloride (TTC) (1% m/V aqueous solution)	10 mL
Distilled water	1 litre

Steam the ingredients to dissolve. The pH should be 7.2 without the need for adjustment. Sterilize TTC by filtration, and pour the medium directly into Petri dishes without further sterilization. The medium should not be stored and re-melted, but poured plates may be kept at 4°C for up to 1 month if placed in a sealed container.

Glucose Azide Broth (Mallman and Seligmann 1950)

Double-strength medium:

Beef extract	9 g
Tryptone	30 g
Glucose	15 g
Sodium chloride	15 g
Sodium azide	400 mg
Bromocresol purple (1.5% m/V ethanolic solution)	2 mL
Distilled water	1 litre

Dissolve the ingredients in water by boiling. If necessary, adjust the pH so that after sterilization, the pH is 7.2 ± 0.1. Prepare single strength medium by diluting the double strength medium with an equal amount of distilled water. Double strength medium (in 10 mL and 50 mL volumes) and single strength medium (in 5 mL volumes) should be dispensed into suitably sized containers. Sterilize by autoclaving at 121°C for 15 minutes.

Kanamycin Aesculin Azide Agar

Tryptone	20 g
Yeast extract	5 g
Sodium chloride	5 g
Sodium citrate	1 g
Aesculin	1 g
Iron(III) ammonium citrate	500 mg
Sodium azide	150 mg
Kanamycin sulphate	20 mg
Agar	12 g
Distilled water	1 litre

Sterilize by autoclaving at 121°C for 15 minutes. Final pH should be 7.0 ± 0.2.

Bile Agar 40% (Cowan 1993)

Ox bile (dehydrated)	40 g*
Serum (sterile)	50 mL
Nutrient agar	1 litre

Melt the nutrient agar, add the ox bile and mix to dissolve. Autoclave at 115°C for 10 minutes. Cool at about 55°C and add the serum aseptically. Mix gently and distribute in bottles, tubes or plates.

*NOTE this is equivalent to 400 mL fresh bile.

Bile Aesculin Azide Agar

Ox bile (dehydrated)	10 g
Tryptone	17 g
Peptone	3 g
Yeast extract	5 g
Sodium chloride	5 g
Sodium azide	150 mg
Iron(III) ammonium citrate	500 mg
Aesculin	1 g
Agar	12 g
Distilled water	1 litre

Steam to dissolve the ingredients and adjust to pH 7.0. Sterilize at 115°C for 10 minutes. Pour into Petri dishes and store at 4°C in sealed containers.

Glucose Phenolphthalein Broth (Clarke 1953)

Basal medium:

Meat extract	10 g
Peptone	10 g
Sodium chloride	5
Distilled water	1 litre

Dissolve the ingredients in the water and adjust the pH to between 7.2–7.4. Sterilize at 115°C for 10 minutes. Prepare a 20% m/V aqueous solution of glucose and sterilize by filtration. Add 50 mL aseptically to 950 mL of the basal medium.

Glycine buffer:

Glycine	600 mg
Sodium chloride	350 mg
Distilled water, freshly boiled	60 mL
0.1M Sodium hydroxide	40 mL

Dissolve the glycine and sodium chloride in the hot water and then add the sodium hydroxide.

Complete medium:

Glucose broth	900 mL
Glycine buffer	100 mL
Phenolphthalein (0.2% m/V aqueous solution)	5 mL

Mix and keep overnight in the refrigerator in a stoppered flask. Sterilize by filtration and distribute aseptically into sterile 5 mL screw-capped bottles leaving as little air space as possible. Incubate at 37°C overnight to check sterility and discard any bottle showing growth or not having a definite pink colour.

Salt Broth (Cowan 1993)
(Nutrient broth with 6.5% m/V sodium chloride)

Meat extract	10 g
Peptone	10 g
Sodium chloride	65 g
Distilled water	1 litre

Dissolve the ingredients in the water, adjust pH to between 7.2–7.4. Distribute in 5 mL volumes and sterilize at 115°C for 10 minutes.

MEDIA FOR SULPHITE-REDUCING CLOSTRIDIA

Tryptose—Sulphite—Cycloserine Agar (Harmon, Kautter and Peeler 1971, Hauschild and Hillsheimer 1974)

Basal medium:

Tryptose	15 g
Soya peptone	5 g
Yeast extract	5 g
Sodium metabisulphite	1 g
Iron(III) ammonium citrate	1 g
Agar	12 g
Distilled water	1 litre

The final pH should be 7.6 ± 0.2. Dispense 100 mL amounts in screw-capped bottles and sterilize by autoclaving at 121°C for 10 minutes. Cool to 50°C for immediate use or store in a cool dark place.

Antibiotic solution:

Prepare a 1% (m/V) solution of D-cycloserine in distilled water, sterilize by filtration and add 4 mL to 100 mL of the cooled molten agar base immediately before pouring plates to give a concentration of 400 μgmL^{-1}. The solution can be kept at 4°C for at least 2 weeks.

Egg yolk emulsion:

Prepare a 50% egg yolk emulsion by mixing the yolk of one fresh egg with an equal volume of sterile 0.9% saline solution. The yolk can be withdrawn from the yolk sac using a 10 mL syringe fitted with a wide bore needle. Add 4 mL of egg yolk emulsion to each bottle of medium as required. Egg yolk emulsion is commercially available.

Oleandomycin—Polymyxin—Sulphadiazine Perfringens Agar (Handford 1974)

Basal medium:

Tryptone	15 g
Yeast extract	5 g
Soya peptone	5 g
Sodium metabisulphite	1 g
Iron(III) ammonium citrate	1 g
Sulphadiazine (10% m/V solution in molar sodium hydroxide)	1 mL
Agar	12 g
Distilled water to	1 litre

The final pH should be 7.6 ± 0.2. Dispense the medium in 100 mL volumes in screw-capped bottles and sterilize by autoclaving at 121°C for 15 minutes. Cool to 50°C for immediate use or store in a cool, dark place.

Antibiotic solutions:

Concentrated stock solutions are prepared as follows:

Oleandomycin: Dissolve 0.5 g of oleandomycin phosphate in 100 mL sterile distilled water.

Polymyxin: Dissolve 500,000 international units (iu) of polymyxin B sulphate in 5 mL of sterile distilled water.

For each solution dilute 100-fold and add 1 mL of each solution to 100 mL molten basal medium cooled to 50°C to give final concentrations of 0.5 μgmL^{-1} oleandomycin phosphate and 10 $iumL^{-1}$ of polymyxin B sulphate. The concentrated solutions can be kept for up to 2 months at 4°C and the diluted solutions for up to one week at the same temperature.

Shahidi-Ferguson Perfringens Agar (Shahidi and Ferguson 1971)

Basal medium:

Tryptose	15 g
Yeast extract	5 g
Iron(III) ammonium citrate	1 g
Sodium metabisulphite	1 g
Polymyxin B sulphate	30000 units
Kanamycin sulphate	12 mg
Agar	12 g
Distilled water	900 mL
Egg yolk emulsion (50% in sterile 0.9% sodium chloride)	100 mL

Sterilize the basal medium by autoclaving in 900 mL volumes at 121°C for 10 minutes. Cool to 50°C and add the egg yolk emulsion. The antibiotics may be added to the basal medium before or after sterilization. The final pH should be 7.6.

Overlay agar:

Use the basal medium without the addition of egg yolk emulsion but made up to 1000 mL with water. Dispense in 100 mL amounts and sterilize by autoclaving at 121°C for 10 minutes.

Differential Reinforced Clostridial Medium (Gibbs and Freame 1965)

Basal medium—single-strength:

Peptone	10.0 g
Meat extract	10.0 g
Sodium acetate (hydrated)	5.0 g
Yeast extract	1.5 g
Soluble starch	1.0 g
Glucose	1.0 g
L(–) Cysteine hydrochloride	500 mg
Distilled water	1 litre

Add the peptone, meat extract, sodium acetate and yeast extract to 800 mL of the water. Dissolve the starch in the remaining 200 mL by first making a cold slurry with a little of the water, boiling the rest and stirring into the paste. Add the glucose and cysteine and dissolve. Adjust the pH to 7.1–7.2, distribute 25 mL volumes in screw-capped bottles and autoclave at 121°C for 15 minutes.

Basal medium—double-strength:

Prepare in the same way using twice the quantities of ingredients; distribute 10 mL volumes in universal screw-capped bottles and 50 mL volumes in 125 mL screw-capped bottles.

Sodium sulphite and iron(III) citrate solutions:

Prepare solutions of 4% (m/V) sodium sulphite (anhydrous) and 7% (m/V) iron(III) citrate in distilled water. Heat the latter to dissolve. Sterilize by filtration. The solutions may be stored at 4°C for up to 14 days.

Final medium:

On the day of use, mix equal volumes of the two solutions. To give final concentrations of 0.04% sodium sulphite and 0.07% iron(III) citrate, add 0.5 mL of the mixture to each 25 mL volume of the single-strength basal medium, freshly steamed and cooled to exclude dissolved oxygen. To each 10 mL and 50 mL volume of double-strength medium, add 0.4 mL and 2.0 mL respectively of the sulphite-iron mixture.

Crossley Milk Medium (Crossley 1941)

Skimmed-milk powder	100 g
Peptone	10 g
Bromocresol purple	100 mg
Distilled water	1 litre

The final pH should be 6.8 ± 0.2. Cream the ingredients with a little distilled water, and gradually dilute to 1 litre with continuous mixing. Dispense in 10 mL volumes and sterilize by autoclaving at 121°C for 5 minutes.

Litmus Milk

To milk prepared from skimmed-milk powder, add sufficient of a 10% (m/V) aqueous solution of litmus to give a blue-purple colour. Distribute in 5 mL volumes in screw-capped bottles and autoclave at 115°C for 10 minutes.

MEDIA FOR COLONY COUNTS

R2A Agar (Reasoner and Geldreich 1985)

Yeast extract	500 mg
Proteose peptone No. 3 or polypeptone	500 mg
Casamino acids	500 mg
Glucose	500 mg
Soluble starch	500 mg
Dipotassium hydrogen phosphate	300 mg
Magnesium sulphate heptahydrate	50 mg
Sodium pyruvate	300 mg
Agar	12 g
Distilled water	1 litre

Adjust pH to 7.2 with solid dipotassium hydrogen phosphate or potassium dihydrogen phosphate before adding agar. Heat to dissolve agar and sterilize at 121°C for 15 minutes.

Yeast Extract Agar

Yeast extract	3 g
Peptone	5 g
Agar	12 g
Distilled water	1 litre

Dissolve the yeast extract and peptone in the water. Adjust the pH to 7.3. Add the agar and steam to dissolve. Distribute in 15 mL amounts in tubes or containers or larger volumes in screw-capped bottles. Autoclave at 115°C for 10 minutes.

Modified King's A Broth (Medium 19, Drake 1966)

Peptone	20.0 g
Ethanol	25 mL
Potassium sulphate, anhydrous	10.0 g
Magnesium chloride, anhydrous	1.4 g
(or magnesium chloride hexahydrate)	(2.9 g)
Cetrimide (cetyltrimethyl ammonium bromide)	0.5 g
Distilled water to	1 litre

Steam to dissolve the ingredients. Distribute in screw-capped bottles and autoclave at 115°C for 10 minutes. Remove the bottles from the autoclave promptly after this sterilization cycle has been completed in order to prevent excessive loss of ethanol from the medium. Alternatively, add filter-sterilized ethanol aseptically to the sterile broth. The final pH should be approximately 7.2; no adjustment is usually necessary.

Asparagine Broth with Ethanol (Medium 10, Drake 1966)

	Single-strength medium	Concentration for use in Repli-dishes
L(–) Asparagine	2 g	3.2 g
L(–) Proline	1 g	1.6 g
Dipotassium hydrogen phosphate, anhydrous	1 g	1.6 g
Magnesium sulphate heptahydrate	500 mg	800 mg
Potassium sulphate	10 g	16.0 g
Ethanol	25 mL	40 mL
Distilled water to	1 litre	1 litre

Steam the ingredients to dissolve. Distribute the single strength medium in 4 mL volumes in screw-capped bijou bottles, and the 40 mL volumes in bottles of 50–100 mL capacity; dispense the concentrated medium in larger bottles for storage. Autoclave at 115°C for 10 minutes. Remove the bottles promptly from the autoclave after this sterilization cycle has been completed in order to prevent excessive loss of ethanol from the medium. Alternatively, add filter-sterilized ethanol aseptically to the sterile medium. The final pH should be approximately 7.2; no adjustment is usually necessary.

Milk Agar with Cetrimide (Brown and Foster 1970)

Yeast extract broth:

Yeast extract	75 mg
Peptone	2500 mg
Sodium chloride	1250 mg
Distilled water	250 mL

Dissolve the ingredients and adjust the pH to between 7.2–7.4. Autoclave at 115°C for 10 minutes.

Final medium:

Skimmed milk powder	100 g
Yeast extract broth	250 mL
Agar	15 g
Cetrimide (Cetyltrimethyl ammonium bromide)	300 mg
Distilled water	750 mL

Add the cetrimide and agar to the yeast extract broth and steam to dissolve. Mix the skimmed milk powder and distilled water. Autoclave both solutions separately at 121°C for 5 minutes and remove them promptly from the autoclave after this sterilization cycle has been completed in order to prevent caramelization of the lactose in the milk. Cool to 50–55°C, mix aseptically and pour into Petri dishes. Store the plates at 4°C for not longer than 4 weeks in sealed containers to prevent drying,

Thermophile-free skimmed milk powder for microbiological purposes is available commercially.

Pseudomonas Agar (based on Drake 1966, Mead and Adams 1977)

Basal medium:

Acid hydrolysed peptone or casein hydrolysate	10.0 g
Gelatine peptone	16.0 g
Potassium sulphate (anhydrous)	10.0 g
Magnesium chloride (anhydrous)	1.4 g
Agar	11.0 g
Distilled water	1 litre

Add 10 mL glycerol to the above ingredients and steam to dissolve; autoclave at 121°C for 15 minutes. Cool to approximately 50°C and add the following filter-sterilized solutions prepared as 1 % aqueous solutions:

(i) Cephaloridine Add 5 mL to 1 litre of basal medium to give a final concentraion of 50 mgL^{-1}:

(ii) Fucidin Add 1 mL to 1 litre of basal medium to give a final concentraion of 10 mgL^{-1}:

(iii) Cetrimide Add 1 mL to 1 litre of basal medium to give a final concentraion of 10 mgL^{-1}. It may be necessary to warm the solution gently to dissolve the cetrimide prior to filtration.

MEDIUM FOR AEROMONAS

Ampicillin-dextrin agar

Tryptose	5 g
Dextrin	10 g
Yeast extract	2 g
Sodium chloride	3 g
Potassium chloride	2 g
Magnesium sulphate heptahydrate	200 mg
Iron (III) chloride	100 mg
Bromothymol blue	80 mg
Agar	15 g
Sodium desoxycholate	100 mg
Ampicillin	10 mg
Distilled water	1 litre

Dissolve all the ingredients except the agar, ampicillin and desoxycholate in the water. Adjust the pH to 8.0. Add the agar and dissolve by boiling. Autoclave at 121°C for 15 minutes. Cool to about 55°C and add the ampicillin (10 mL of a freshly prepared solution containing 1 mgmL^{-1}) and desoxycholate (10 mL of a solution of 10 mgmL^{-1}). Mix well and pour plates. Keep plates in a refrigerator and use within 7 days. For marine or estuarine samples 0.05 g of 2,4-diamino-6,7-diisopropyl pteridine phosphate (O/129 phosphate) can be added to each litre of medium before autoclaving.

MEDIA FOR THERMOPHILIC CAMPYLOBACTERS

Preston Enrichment Broth (Bolton and Robertson 1982, Bolton et al 1983)

Basal medium:

Nutrient broth No 2	25 g
Iron(II) sulphate	250 mg
Sodium metabisulphite	250 mg
Sodium pyruvate	250 mg
Distilled water	950 mL

Dissolve and sterilize by autoclaving at 121°C for 15 minutes.

Blood supplement:

Add 50 mL saponin-lysed horse blood to the basal medium to give a final concentration of 5%.

Antibiotic solutions:

The following should be filter-sterilized:

(i) Trimethoprim: Dissolve 0.5 g trimethoprim lactate in 100 mL distilled water. Add 2 mL of this solution to the basal medium to give a final concentration of 10 mgL^{-1};

(ii) Polymyxin: Dissolve 500,000 units of polymyxin B sulphate in 40 mL distilled water. Add 0.4 mL to the basal medium to give a final concentration of 5000 units L^{-1};

(iii) Rifampicin: Dissolve 0.2 g of rifampicin in 20 mL methanol. Add 1 mL to the basal medium to give a final concentration of 10 mgL^{-1};

(iv) Actidione: Dissolve 2 g of actidione in 10 mL of methanol and add 10 mL sterile distilled water. Add 1 mL to the basal medium to give a final concentration of 100 mgL^{-1}.

Preston Selective Agar (Bolton et al 1982)

Basal medium:

Nutrient broth No. 2	25 g
Agar	12 g
Distilled water	950 mL

Sterilize by autoclaving at 121°C for 15 minutes.

Antibiotic solution and blood supplements:

Allow the basal medium to cool to approximately 50°C and add the antibiotics and blood supplement as described for the Preston Enrichment Broth.

Blood-Free Selective Agar (Bolton, Hutchinson and Coates 1986)

Nutrient broth No. 2	25 g
Bacteriological charcoal	4 g
Casein hydrolysates	3 g
Sodium desoxycholate	1 g
Iron(II) sulphate	250 mg
Sodium pyruvate	250 mg
Agar	12 g
Distilled water	1 litre

Adjust pH to 7.4 and sterilize by autoclaving at 121°C for 15 minutes. Cool to approximately 50°C and add (i) 1 mL of a filter-sterilized aqueous solution containing 32000 mgL^{-1} of cefoperazone, to give a final concentration of 32 mgL^{-1}; (ii) 1 mL of a filter-sterilized aqueous solution containing 10000 mgL^{-1} of amphotericin to give a final concentration of 10 mgL^{-1}.

MEDIA FOR *E. COLI* O157:H7

Modified buffered peptone water (Chapman et al 1993)

Peptone	10.0 g
Sodium chloride	5.0 g
Disodium hydrogen phosphate	3.5 g
Potassium dihydrogen phosphate	1.5 g
Distilled water	1 litre

Dissolve the ingredients in the water. Distribute in suitable volumes in screw-capped bottles and autoclave at 121°C for 15 minutes. The final pH should be 7.2 ± 0.1. Usually, no adjustment is necessary.

Antibiotic solution:

The following can be purchased as sterile powders and dissolved in sterile water, or filter-sterilized.

(i) Vancomycin: Dissolve 80 mg of vancomycin hydrochloride in 10 mL distilled water. The solution may be stored at –20°C and used within 3 months. Add 1 mL of the solution to the basal medium to give a final concentration of 8 mgL^{-1}.

(ii) Cefixime: Dissolve 0.5 g of cefixime in 100 mL ethanol. Store at 4°C and use within 3 months. Dilute 1 mL to 100 mL in ethanol and add 1 mL to the basal medium to give a final concentration of 0.05 mgL^{-1}.

(iii) Cefsulodin: Dissolve 100 mg of cefsulodin (sodium salt) in 10 mL of distilled water. The solution may be stored at –20°C and should be used within 3 months. Add 1 mL of this solution to the basal medium to five a final concentration of 10 mgL^{-1}.

Modified Sorbitol MacConkey Agar (Zadik et al 1993)

Peptone	20.0 g
Sorbitol	10.0 g
Bile salts No 3	1.5 g
Sodium chloride	5.0 g
Neutral red	30 mg
Crystal violet	1 mg
Agar	15.0 g
Distilled water	1 litre

Add the ingredients to the water and steam to dissolve. Distribute in suitable volumes into screw-capped bottles and sterilize by autoclaving at 121°C for 15 minutes. Allow to cool and store in the dark at room temperature. The solutions can be stored for up to 1 month before use. Alternatively, cool to 50°C and add the selective supplements. Pour into sterile 55 or 90 mm Petri dishes and allow to set. Plates may be stored for up to 1 month at 4°C and should be dried in a suitable oven before use.

Selective supplements

The following can be purchased as sterile powders and dissolved in sterile water or filter-sterilized.

(i) Cefixime: Dissolve 0.5 g of cefixime in 100 mL ethanol. Store at 4°C and use within 3 months. Dilute 1 mL to 100 mL in ethanol and add 1 mL to the basal medium to give a final concentration of 0.05 mgL^{-1}.

(ii) Potassium tellurite: Dissolve 25 mg of potassium tellurite in 10 mL distilled water. The solution may be stored at –20°C and should be used within 3 months. Add 1 mL of this solution to the basal medium to give a final concentration of 2.5 mgL^{-1}.

MEDIA FOR SALMONELLAS

Buffered Peptone Water (Edel and Kampelmacher 1973)

Peptone	10.0 g
Sodium chloride	5.0 g
Disodium hydrogen phosphate	3.5 g
Potassium dihydrogen phosphate	1.5 g
Distilled water	1 litre

Dissolve the ingredients. Distribute in bulk quantities in screw-capped bottles and autoclave at 121 °C for 15 minutes. The final pH should be approximately 7.2. Usually, no adjustment is needed.

Rappaport-Vassiliadis Enrichment Broth (based on Vassiliadis et al 1976, Vassiliadis 1983)

Solution A:

Tryptone	5.0 g
Sodium chloride	8.0 g
Potassium dihydrogen phosphate	1.6 g
Distilled water	1 litre

Dissolve the ingredients in the distilled water by boiling on the day of use.

Solution B:

Magnesium chloride hexahydrate	40.0 g
Distilled water	100 mL

Solution C:

Malachite green	0.4 g
Distilled water	100 mL

For use, add 100 mL of Solution B and 10 mL of Solution C to 1 litre of freshly prepared Solution A. Dispense in 10 mL volumes and sterilize by autoclaving at 115°C for 10 minutes. The final pH should be approximately 5.2.

Selenite Broth (Hobbs and Allison 1945a,b)

Tryptone or Polypeptone	5 g
Mannitol	4 g
Disodium hydrogen phosphate	10 g
Sodium hydrogen selenite	4 g
Distilled water	1 litre

The final pH should be approximately 7.0. If the medium is to be used immediately, warm gently to dissolve the ingredients. If stored before use then sterilize either by filtration or by steaming for no more than 30 minutes.

The final medium should have a pale straw colour. Overheating produces a red precipitate of selenium with a consequent loss of selectivity and the medium should be discarded.

Brilliant Green Agar (ICMSF 1978)

Yeast extract	3.0 g
Proteose peptone or polypeptone	10.0 g
Sodium chloride	5.0 g
Lactose	10.0 g
Sucrose	10.0 g
Phenol red (0.2% m/V aqueous solution)	40.0 mL
Brilliant green (0.5% m/V aqueous solution)	2.5 mL
Agar	20.0 g
Distilled water	960 mL

Steam to dissolve the ingredients and dispense in appropriate volumes. Sterilize by autoclaving at 121°C for 15 minutes. The final pH after sterilization should be 6.9 ± 0.1.

Bismuth Sulphite Agar (Modified from Wilson and Blair 1927)

Wilson and Blair stock agar:

Peptone	10 g
Beef extract	6 g
Iron(III) citrate	400 mg
Brilliant green (1% m/V aqueous solution)	1 mL
Agar	20 g
Distilled water	1 litre

Steam to dissolve the ingredients and sterilize by autoclaving at 121°C for 15 minutes. Cool to approximately 50°C and add solution B as described below.

Wilson and Blair solution B:

Bismuth ammonium citrate	3 g
Sodium sulphite	5 g
Glucose	5 g
Disodium hydrogen phosphate	5 g
Distilled water	40 mL

Dissolve the ingredients by gently bringing to the boil. The colour of the solution should be grey-white. Any blackening of the solution indicates overheating and the solution should be discarded.

Cool rapidly in cold water and add 40 mL to 1 litre of basal medium.

After pouring, the plates should be held for two days in a refrigerator to oxidise prior to use.

Xylose Lysine Desoxycholate Agar (Taylor and Harris 1965)

Basal medium:

Lactose	7.5 g
Sucrose	7.5 g
Xylose	3750 mg
L(–)Lysine hydrochloride	5.0 g
Sodium chloride	5.0 g
Yeast extract	3.0 g
Phenol red (0.4% m/V aqueous solution)	20 mL
Agar	12.0 g
Distilled water	1 litre

Steam to dissolve ingredients and distribute in screw-capped bottles. Autoclave at 115°C for 10 minutes.

Solution A:

Sodium thiosulphate pentahydrate	34 g
Iron(III) ammonium citrate	4 g
Distilled water	100 mL

Heat gently to dissolve. Pasteurize at 60°C for 1 hour.

Solution B:

Sodium desoxycholate	10 g
Distilled water	100 mL

Dissolve and pasteurize at 60°C for 1 hour.

Final medium:

For use, melt the basal medium, cool to approximately 50°C and add aseptically 2.0 mL of Solution A per 100 mL of basal medium. Mix gently. With a separate pipette, add aseptically 2.5 mL of Solution B per 100 mL of basal medium. Mix and pour into Petri dishes. The final pH should be 7.3.

Lysine Iron Agar (Edwards and Fife 1961)

Peptone	5 g
Yeast extract	3 g
Glucose	1 g
L(–)Lysine	10 g
Iron(III) ammonium citrate	500 mg
Sodium thiosulphate pentahydrate	40 mg
Bromocresol purple	20 mg
Agar	15 g
Distilled water	1 litre

Steam to dissolve ingredients and dispense in 5 mL volumes in small test tubes. Autoclave at 115°C for 10 minutes. Cool in a sloping position to give agar slopes with a deep butt. The final pH should be approximately 6.7. No adjustment is usually necessary.

Triple Sugar Iron Agar (IAMS 1958)

Lab-Lemco (beef extract)	3 g
Yeast extract	3 g
Peptone	20 g
Sodium chloride	5 g
Lactose	10 g
Sucrose	10 g
Glucose	1 g
Iron(III) citrate	300 mg
Sodium thiosulphate pentahydrate	300 mg
Phenol red (0.4% m/V aqueous solution)	6 mL
Agar	15 g
Distilled water to	1 litre

Steam to dissolve ingredients. Add the indicator. Dispense in 5 mL volumes into small test tubes and autoclave at 115°C for 10 minutes. Cool in a sloping position to form a slope with a deep butt. The final pH should be approximately 7.4. No adjustment is usually necessary.

Urea Broth

Broth base:

Peptone	1 g
Glucose	1 g
Disodium hydrogen phosphate	1 g
Potassium dihydrogen phosphate	800 mg
Sodium chloride	5 g
Phenol red (0.4% m/V aqueous solution)	1 mL
Distilled water	1 litre

Dissolve the ingredients and check the pH is approximately 6.8. Dispense in 95 mL volumes in screw-capped bottles and autoclave at 115°C for 10 minutes.

For use, add aseptically 5 mL of 40% (m/V) solution of urea sterilized by filtration to 95 mL Broth Base. Dispense aseptically in 2–3 mL volumes in sterile bijou bottles.

MEDIA FOR SHIGELLAS

Modified Hajna GN Enrichment Broth

Tryptone	20 g
Glucose	1 g
D-Mannitol	2 g
Sodium citrate	5 g
Sodium desoxycholate	500 mg
Dipotassium hydrogen phosphate	4 g
Potassium dihydrogen phosphate	1500 mg
Sodium chloride	5 g
DL serine	1 g
Distilled water	1 litre

The ingredients are heated to dissolve, distributed into universal bottles in 10 mL amounts and steamed (100°C) for 30 minutes. The final pH should be 7.2.

Hektoen Agar (King and Metzger 1968)

Yeast extract	3 g
Proteose peptone	12 g
Lactose	12 g
Sucrose	12 g
Salicin	2 g
Iron(III) ammonium citrate	1500 mg
Acid fuchsin	100 mg
Bromothymol blue	65 mg
Bile salts No 3	9 g
Sodium chloride	5 g
Sodium thiosulphate pentahydrate	5 g
Agar	14 g
Novobiocin	15 µg
Distilled water	1 litre

Suspend the ingredients in 1 litre of distilled water and soak for 10 minutes. Heat gently and allow to boil for a few seconds to dissolve the agar. Do not autoclave. Cool to 60°C and pour plates.

Modified Desoxycholate Citrate Agar

Tryptone	20.0 g
Lactose	10.0 g
Sodium thiosulphate pentahydrate	6.8 g
Iron(III) ammonium citrate	800 mg
Neutral red	30 mg
Sodium desoxycholate	500 mg
DL serine	1.0 g
Tetracycline hydrochloride	32 mg
Agar	14.0 g
Distilled water	1 litre

The ingredients are dissolved by heating gently until boiling. The medium is then cooled to 50°C and the tetracycline hydrochloride (which has been separately sterilized by filtration through a 0.2 µm membrane filter) is added to give a final concentration of 32 mgL^{-1}. The medium is then poured into sterile Petri dishes.

MEDIUM FOR STAPHYLOCOCCI

Modified lipovitellin-salt-mannitol agar (Stengren and Starzyk 1984)

Beef extract	1.0 g
Proteose peptone	10.0 g
D(–)mannitol	10.0 g
Sodium chloride	75.0 g
Agar	15.0 g
Phenol red	25 mg
Sodium azide	50 mg
Sterile egg yolk	20 mL
Distilled water	1 litre

Dissolve all the ingredients except the egg yolk by boiling. Adjust the pH, if necessary, so that after sterilization the pH is 7.4 ± 0.2. Sterilize by autoclaving at 121°C for 15 minutes, allow to cool to 50°C and then aseptically add the egg yolk. Mix to distribute the egg yolk evenly and distribute in Petri dishes. If the medium is only to be used for membrane filtration the egg yolk may be omitted.

MEDIA FOR VIBRIOS

Alkaline peptone water

Peptone	10 g
Sodium chloride	5 g
Distilled water	1 litre

Adjust pH to 8.6 with molar sodium hydroxide solution. Sterilize by autoclaving at 121°C for 15 minutes.

Thiosulphate Citrate Bile Salt Sucrose Agar (Kobayashi et al 1963)

Yeast extract	5 g
Peptone	10 g
Sodium thiosulphate pentahydrate	10 g
Sodium citrate	10 g
Ox bile	8 g
Sucrose	20 g
Sodium chloride	10 g
Iron(III) citrate	1 g
Bromothymol blue	40 mg
Thymol blue	40 mg
Agar	14 g
Distilled water	1 litre

Suspend the ingredients in distilled water and dissolve by boiling. Do not autoclave. The final pH should be 8.6 ± 0.2.

MEDIA FOR YERSINIAS

Tris-buffered peptone

Peptone	10.0 g
Tris-(hydroxymethyl) aminomethane	12.1 g
Sodium chloride	5.0 g
Distilled water	1 litre

Dissolve the ingredients in the water. Adjust pH to 8.0 with 10M hydrochloric acid. Sterilize by autoclaving at 121°C for 15 minutes.

Potassium hydroxide/sodium chloride solution

Potassium hydroxide	500 mg
Sodium chloride	500 mg

Dissolve in distilled water and make up to 100 mL.

Cefsulodin Irgasan Novobiocin Agar (Mair and Fox 1986)

Special peptone	20.0 g
Yeast extract	2.0 g
D(–)Mannitol	20.0 g
Sodium pyruvate	2.0 g
Sodium chloride	1.0 g
Magnesium sulphate	10 mg
Sodium desoxycholate	500 mg
Neutral red	30 mg
Crystal violet	1 mg
Agar	12.5 g
Distilled water	1 litre

Antibiotic supplement:

Cefsulodin	7.50 mg
Irgasan	2.00 mg
Novobiocin	1.25 mg

Dissolve 29 g of base in 500 mL of distilled water by bringing slowly to the boil. Sterilize by autoclaving at 121°C for 15 minutes and cool to 50°C. Re-constitute antibiotic supplement by adding 2 mL of sterile distilled water and 1 mL ethanol to the above ingredients, mix gently and add aseptically to 500 mL molten agar. Pour plates, allow to dry prior to use.

REAGENTS

Kovacs' Reagent (Kovacs 1928)

p-Dimethylaminobenzaldehyde	5 g
Amyl alcohol (analytical grade reagent, free from organic bases)	75 mL
Hydrochloric acid (concentrated)	25 mL

Dissolve the aldehyde in the alcohol. Add the concentrated acid with care. Protect from light and store at 4°C.

NOTE: The reagent should be pale yellow to light brown in colour; some samples of amyl alcohol are unsatisfactory, and give a dark colour with the aldehyde.

Oxidase Reagent

Tetramethyl-p-phenylenediamine hydrochloride	0.1 g
Distilled water	10 mL

This reagent does not keep and it must therefore be freshly prepared for use in small amounts each time that it is needed.

Ringer's Solution (Quarter-strength)

Sodium chloride	2.25 g
Potassium chloride	105 mg
Calcium chloride, anhydrous	120 mg
Sodium bicarbonate	50 mg
Distilled water	1 litre

Dissolve the ingredients and dispense in convenient volumes. Sterilize by autoclaving at 121°C for 15 minutes.

Filter-Aid (Hammarstrom and Ljutove 1954)

Hyflo-supercel	1 g
Distilled water	15 mL

Add the filter-aid to the water in a screw-capped universal bottle and autoclave at 121°C for 15 minutes. Prepare several bottles at the same time and store until required.

Appendix C: Tables of Most Probable Numbers

These tables indicate, from the various combinations of positive and negative reactions for the different volumes examined, the estimated number of bacteria in 100 mL of sample. It is important to realise that this, the most probable number (MPN), is only an estimate based on statistical probability and that the true number may lie within a range of values (see section 6.3.2.3). Approximate 95% confidence intervals, which demonstrate a range of possible counts which could yield the tabulated number of positive tubes, have been published (Swaroop 1951). A computing procedure for estimating these approximate intervals for other dilution series has been published (Hurley and Roscoe 1983). These intervals can be explored more accurately (Tillett and Coleman 1985) but are seldom of practical use when reporting results, because they apply to the accuracy of the method and not the likely range of organisms at the sampling source. The tables in this Appendix give the MPN and the MPR as described in section 6.3.2.3. The ranges of numbers are close contenders for the title MPN and each count has a probability of being the true count which is at least 0.95 times as large as the probability that the MPN is correct. The MPR illustrates situations where the method becomes imprecise, particularly when nearly all the tubes show growth and, in practice, a further dilution should have been made to give a clearer estimate of the MPN.

Tables 10 and 11 give the MPN (and where applicable the MPR) for a 6-tube series containing 1 × 50 mL and 5 × 10 mL volumes; and an 11-tube series with 1 × 50 mL, 5 × 10 mL and 5 × 1 mL volumes respectively. Table 12 relates to a 15-tube series with 5 × 10 mL, 5 × 1 mL and 5 × 0.1 mL volumes, but only the likely combinations of positive and negative reactions are listed. For example, multiple positive reactions in the 0.1 mL tubes would not be expected if all the 10 mL and 1 mL tubes are negative and so combinations of 0,0,2 etc are not tabulated. If these unlikely combinations are observed in practice with a frequency greater than about 1 in 100 tests, it is an indication that the statistical assumptions underlying the MPN estimation are not being fulfilled (Swaroop 1951, Woodward 1957, Man 1975). For example, the organisms may not have been uniformly distributed throughout the sample, or toxic substances may have been present.

Calculation of MPN Values

Record the number of positive reactions for each set of tubes and, from the relevant table, read the MPN of organisms present in 100 mL of the sample.

Where a series of dilutions of the sample is used, apply the following rules:

(i) Use only three consecutive sets of dilutions for calculating the MPN.

(ii) Select, wherever possible, three consecutive dilutions where the results are neither all positive nor all negative. The most efficient statistical estimate will result when about half the tubes are positive. (See examples a, b and c in Table 9).

Table 9 Examples of the derivation of the MPN from the numbers of positive reactions in a series of dilutions.*

Example in text	Volume of sample (mL)					MPN per 100 mL
	10	1	0.1	0.01	0.001	
a	**5**	**3**	**2**	0		140
b	5	**5**	**3**	**2**	0	1400
c	5	**5**	**2**	**0**	0	500
d	**3**	**1**	**0**	0		9
e	**0**	**1**	**0**	0		2

*Numbers in bold type indicate which results should be used in determining the MPN values.

(iii) If less than three sets of dilutions give positive results, start with the set containing the largest volume of the sample (see example d in Table 9).

(iv) If only one set of the tubes gives a positive reaction, use this dilution and the one higher and one lower (see example e in Table 9).

Table 10 MPN and MPR per 100 mL of sample for a 6-tube series containing 1 × 50 mL and 5 × 10 mL volumes.

Number of tubes giving a positive reaction		MPN per 100 mL	MPR* per 100 mL
1 × 50 mL	5 × 10 mL		
0	0	none found	
0	1	1	
0	2	2	
0	3	3	
0	4	4	4–5
0	5	6	
1	0	1	
1	1	2	
1	2	5	4–5
1	3	9	8–10
1	4	15	13–18
1	5	>18**	–

*MPR gives counts which are at least 95% as probable as the MPN in being the correct number (see sections 6.3.2.3 and 7.6.2)
**There is no discrimination when all tubes show growth; the theoretical MPN is infinity. The true count is likely to exceed 18.

Table 11 MPN and MPR per 100 mL of sample for an 11-tube series of 1 × 50 mL, 5 × 10 mL and 5 × 1 mL volumes.

Number of tubes giving a positive reaction			MPN per 100 mL	MPR* per 100 mL
1 × 50 mL	5 × 10 mL	5 × 1 mL		
0	0	0	none found	
0	0	1	1	
0	1	0	1	
0	1	1	2	
0	2	0	2	
0	2	1	3	
0	3	0	3	
1	0	0	1	
1	0	1	2	
1	1	0	2	
1	1	1	4	
1	1	2	6	
1	2	0	4	4–5
1	2	1	7	6–7
1	2	2	9	9–10
1	3	0	8	7–9
1	3	1	10	10–11
1	3	2	13	12–15
1	3	3	17	15–18
1	4	0	12	11–14
1	4	1	16	15–19
1	4	2	21	19–24
1	4	3	27	24–30
1	4	4	33	30–38
1	5	0	23	20–27

(**Table 11** continued)

Number of tubes giving a positive reaction			MPN per 100 mL	MPR* per 100 mL
1 × 50 mL	5 × 10 mL	5 × 1 mL		
1	5	1	33	29–40
1	5	2	53	44–65
1	5	3	91	75–110
1	5	4	160	134–190
1	5	5	>180**	

*MPR gives counts which are at least 95% as probable as the MPN in being the correct number (see sections 6.3.2.3 and 7.6.2)
**There is no discrimination when all tubes show growth; the theoretical MPN is infinity. The true count is likely to exceed 180.

Table 12 MPN and MPR per 100 mL of sample for a 15-tube series containing 5 × 10 mL, 5 × 1 mL and 5 × 0.1 mL test volumes.

Number of tubes giving a positive reaction			MPN per 100 mL	MPR* per 100 mL
5 × 10 mL	5 × 1 mL	5 × 0.1 mL		
0	0	0	none found	
0	0	1	2	
0	1	0	2	
1	0	0	2	
1	0	1	4	
1	1	0	4	
1	2	0	5	
2	0	0	4	
2	0	1	5	
2	1	0	5	
2	1	1	7	
2	2	0	7	7–9
2	3	0	11	
3	0	0	7	
3	0	1	9	
3	1	0	9	
3	1	1	13	
3	2	0	13	
3	2	1	16	14–16
3	3	0	16	14–16
4	0	0	11	11–13
4	0	1	14	14–16
4	1	0	16	14–16
4	1	1	20	18–20
4	2	0	20	18–22
4	2	1	25	23–27
4	3	0	25	23–27
4	3	1	31	29–34
4	4	0	32	29–34
4	4	1	38	34–41
5	0	0	22	20–23
5	0	1	29	25–34
5	0	2	41	36–50
5	1	0	31	27–36
5	1	1	43	36–50
5	1	2	60	50–70
5	1	3	85	70–95
5	2	0	50	40–55
5	2	1	70	60–80

(**Table 12** continued)

Number of tubes giving a positive reaction			MPN per 100 mL	MPR* per 100 mL
5 × 10 mL	5 × 1 mL	5 × 0.1 mL		
5	2	2	95	80–110
5	2	3	120	105–135
5	3	0	75	65–90
5	3	1	110	90–125
5	3	2	140	120–160
5	3	3	175	155–200
5	3	4	210	185–240
5	4	0	130	110–150
5	4	1	170	150–200
5	4	2	220	190–250
5	4	3	280	240–320
5	4	4	345	300–390
5	5	0	240	200–280
5	5	1	350	290–420
5	5	2	540	450–600
5	5	3	910	750–1100
5	5	4	1600	1350–1900
5	5	5	>1800**	

*MPR gives counts which are at least 95% as probable as the MPN in being the correct number (see sections 6.3.2.3 and 7.6.2)
**There is no discrimination when all tubes show growth; the theoretical MPN is infinity. The true count is likely to exceed 1800.

Table 13 95% Confidence Intervals (95% CI) for the (unobserved) count from the second half-sample for the observed count from the first half-sample of 0 to 100 (see section 6.2.6)

Observed count in first half-sample	95% CI for unobserved count in second half-sample	Observed count in first half-sample	95% CI for unobserved count in second half-sample
0	0–5	24	12–40
1	0–7	25	13–41
2	0–9	26	13–43
3	0–11	27	14–44
4	0–12	28	15–45
5	0–14	29	16–47
6	1–16	30	16–48
7	1–17	31	17–49
8	2–19	32	18–50
9	2–20	33	19–52
10	3–22	34	19–53
11	3–23	35	20–54
12	4–24	36	21–55
13	5–26	37	22–56
14	5–27	38	22–58
15	6–28	39	23–59
16	6–30	40	24–60
17	7–31	41	25–61
18	8–32	42	26–63
19	8–34	43	26–64
20	9–35	44	27–65
21	10–36	45	28–66
22	10–38	46	29–67
23	11–39	47	29–69

(**Table 13** continued)

Observed count in first half-sample	95% CI for unobserved count in second half-sample	Observed count in first half-sample	95% CI for unobserved count in second half-sample
48	30–70	75	52–102
49	31–71	76	53–103
50	32–72	77	54–104
51	33–73	78	55–105
52	33–75	79	56–106
53	34–76	80	57–107
54	35–77	81	58–108
55	36–78	82	58–110
56	37–79	83	59–111
57	38–80	84	60–112
58	38–82	85	61–113
59	39–83	86	62–114
60	40–84	87	63–115
61	41–85	88	63–117
62	42–86	89	64–118
63	42–88	90	65–119
64	43–89	91	66–120
65	44–90	92	67–121
66	45–91	93	68–122
67	46–92	94	69–123
68	47–93	95	69–125
69	47–95	96	70–126
70	48–96	97	71–127
71	49–97	98	72–128
72	50–98	99	73–129
73	51–99	100	74–130
74	52–100		

Appendix D: List of Abbreviations used in this Report:

Organisations and personnel	AVAB	Automatic Vending Association of Britain
	CCDC	Consultant in Communicable Disease Control
	CDSC	Communicable Disease Surveillance Centre (PHLS)
	CD(S)U	Communicable Diseases (Scotland) Unit
	DH	Department of Health
	DHSS	Department of Health and Social Security
	DoE	Department of the Environment
	DPH	Director of Public Health
	DT	Department of Transport
	DWI	Drinking Water Inspectorate
	EC	European Community
	EHO	Environmental Health Officer
	ICD	Infection Control Doctor
	MAFF	Ministry of Agriculture, Fisheries and Food
	MOEH	Medical Officer for Environmental Health
	NHS	National Health Service
	NRA	National Rivers Authority
	NWC	National Water Council
	PHLS	Public Health Laboratory Service
	SCA	Standing Committee of Analysts (DoE)
	SHHD	Scottish Home and Health Department
	WAA	Water Authorities Association
	WHO	World Health Organization
	WO	Welsh Office
	WOAD	Welsh Office Agriculture Department
Media, reagents etc	ABE	Asparagine Broth and Ethanol
	ADA	Ampicillin Dextrin Agar
	AODC	Acridine Orange Direct Count
	BAA	Bile Aesculin Agar
	BiA	Bile Agar
	BlA	Blood Agar
	BGA	Brilliant Green Agar
	BGLBB	Brilliant Green Lactose Bile Broth
	BPW	Buffered Peptone Water
	BSA	Bismuth Sulphite Agar
	CI	Confidence Interval
	CINA	Cefsulodin Irgosan Novobiocin Agar
	CFU	Colony Forming Unit
	CMM	Crossley's Milk Medium
	DRCM	Differential Reinforced Clostridial Medium
	GAB	Glucose Azide Broth
	GPB	Glucose Phenolphthalein Broth
	HA	Hektoen Agar
	HEB	Hajna GN Enrichment Broth
	KAAA	Kanamycin Aesculin Azide Agar
	LIAS	Lysine Iron Agar Slope
	LMM	Litmus Milk Medium
	LPW	Lactose Peptone Water
	LTLB	Lauryl Tryptose Lactose Broth
	M-5LSMA	Modified Lipovitellin-Salt-Mannitol Agar
	MA	MacConkey Agar
	MAC	Milk Agar with Cetrimide
	MBPW	Modified Buffered Peptone Water

MDCA	Modified Desoxycholate Citrate Agar
MEA	Membrane Enterococcus Agar
MKB	Modified King's A Broth
MLSB	Membrane Lauryl Sulphate Broth
MMGM	Minerals Modified Glutamate Medium
MPN	Most Probable Number
MPR	Most Probable Range
MSMA	Modified Sorbitol MacConkey Agar
NA	Nutrient Agar
NB	Nutrient Broth
OPSPA	Oleandomycin Polymyxin Sulphadiazine Perfringens Agar
PA	*Pseudomonas* Agar
P-A	Presence-Absence
PEB	Preston Enrichment Broth
PSA	Preston Selective Agar
PSO	*Salmonella* Polyvalent O Antiserum
PSH	*Salmonella* Polyvalent H Antiserum
R2A	Reasoner's No 2 Agar
RVB	Rappaport Vassiliadis Broth
SB	Selenite Broth
SFPA	Shahidi-Ferguson Perfringens Agar
TCBSA	Thiosulphate Citrate Bile Salt Sucrose Agar
TSCA	Tryptose Sulphite Cycloserine Agar
TSIAS	Triple Sugar Iron Agar Slope
TTC	2,3,5-triphenyltetrazolium chloride
TW	Tryptone Water
UB	Urea Broth
XLDA	Xylose Lysine Desoxycholate Agar
YEA	Yeast Extract Agar

Appendix E: Legislation and Related References

The Act	The Water Industry Act 1991. In Scotland: The Water (Scotland) Act 1980 as amended by Schedule 22 of the Water Act 1989.
The COSHH regulations	The Control of Substances Hazardous to Health Regulations 1988 (SI 1988/1657).
The Directive	The European Community Directive relating to the quality of water intended for human consumption (Council Directive 80/778/EEC).
The Food Act	The Food Safety Act 1990 (See also the Drinking Water in Containers Regulations 1994 SI 1994/743) and the Natural Mineral Waters Regulations 1985(SI 1985/71 as amended by SI 1990/2487.
The Guidance Document	'Guidance on safeguarding the quality of public water supplies' 1989 HMSO ISBN 0 11 752262 7. In Scotland: 'Guidance on safeguarding the quality of public water supplies in Scotland', Scottish Office Environment Department 1990.
The Health and Safety Act	The Health and Safety at Work etc Act 1974.
The Private Supplies Regulations	The Private Water Supplies Regulations 1991 (SI 1991/2790). In Scotland: The Private Water Supplies (Scotland) Regulations 1992 (Scottish Office Circular 20/1992).
The Regulations	The Water Supply (Water Quality) Regulations 1989 (SI 1989/1147) as amended by SI 1989/1384 and SI 1991/1837. In Scotland: The Water Supply (Water Quality) (Scotland) Regulations 1990 No 119 (S.11) as amended by SI 1991/1333 (S.129).

Appendix F: Animalcules in mains water

General introduction

Complaints directly attributable to animals may be due either to

(i) the presence of animals in tap water; or

(ii) discoloured water, caused by animal debris.

One of the earliest records of a mains infestation by animalcules was due to the 'water louse', *Asellus aquaticus*. *Asellus* still causes the greatest number of consumer complaints, especially in water derived from lowland surface waters. This type of source will support higher crops of algae and a greater density of the eggs and larvae of invertebrates which may, if not arrested at the treatment works, enter the distribution system. Upland sources, in contrast, are usually very low in nutrients and consequently support fewer animalcules. Because of this improved quality however, treatment is often minimal, resulting in a greater proportion of those eggs and larvae which are present passing into distribution. Groundwaters are generally the least productive, with an associated distinct invertebrate fauna adapted to the permanent darkness of subterranean life.

Passage of eggs and larvae through the water treatment works is likely to be the major method of entry into the distribution system for most species. Service reservoirs may allow access by aerial insects which then lay their eggs in the water. Ingress into service reservoirs by terrestrial animals such as earthworms, slugs and beetles may occur through cracks, keyholes in covers etc and these types of animals may also enter the mains when being repaired, or when newly laid.

Following entry into the distribution system, however, only a few species are able to colonise and form breeding populations. The most important of these are the water louse *Asellus*, *Gammarus* species and *Crangonyx pseudogracilis* (the 'freshwater shrimps'), *Nais* species (very small, white worms) and some of the small crustaceans such as certain species of the copepod *Cyclops* and the cladoceran *Alona*.

None of the insects are able to complete their life cycle within the mains, with the exception of the unusual, fortunately rare, species of chironomid midges which are able to reproduce parthenogenetically. Some of the chironomid larvae nevertheless live for over two years before pupation, so constant recruitment through a treatment works may allow large numbers to colonise the mains.

Terrestrial animals, clearly, will not survive, but very many aquatic organisms will also be unable to adapt to the conditions in the distribution system; high flows probably being the most common adverse factor. The greatest populations of colonisers are often found in quiescent areas such as at the ends of distribution networks. Most are detritivores, feeding on organic debris; others are carnivorous. A few are filter-feeding organisms, for example the bivalve molluscs *Pisidium* species and these require a constant supply of organic material. Herbivorous leeches may become attached to the walls of the mains by means of their suckers, whilst the tiny Jenkins Spire Shell mollusc, *Potamopyrgus jenkinsi* is able to move amongst the encrustations provided the flow is not too strong. This species is an excellent coloniser, producing numerous live young without the need for a mate, hence an infestation may occur from just one individual.

A wide diversity of animalcules may potentially arrive at consumers' taps—over 150 species have been recorded in water mains. Of these, only a fraction have ever been reported as a consumer complaint, principally because of the small size of most species coupled with a low population density. Those which do cause attention are either large and conspicuous or numerous, small and mobile.

Drowned terrestrial species and larger aquatic animals such as leeches, beetles and certain snails are clearly of concern to a consumer. Such occurrences tend to be occasional in frequency, because of the scarcity of likely points of entry for the organisms. *Asellus* and the gammarids (*Gammarus* and *Crangonyx*) are also large but may be present in greater numbers. Other organisms commonly causing complaint are the mobile worms of the *Nais* genus and *Cyclops* copepods.

Although their presence may indicate a water supply problem, there is no known health risk associated with the animalcules commonly found in mains in the United Kingdom, though in tropical countries intermediate stages of parasites (for example of *Dracunculus medinensis* in *Cyclops*) may be a problem.

It is possible that nematode worms may host pathogenic bacteria in their gut, where they may remain unaffected by chlorine, but there is no evidence to suggest that this presents a public health risk. However, the presence of any of these animalcules may reflect problems with filters at the treatment works or breaches in the integrity of the distribution system, and their presence must lead to an immediate assessment of the microbiological quality of the supply.

Individual animalcules

Asellus species (usually *A. aquaticus*)

These are large (up to 14 mm) isopod crustaceans, dorso-ventrally compressed and superficially resembling the terrestrial woodlouse. Most *Asellus* gain access in the form of eggs via the treatment works and they readily infest mains, clinging to the walls in search of detrital material for food. They are able to breed, twice annually, in the pipework, producing up to 100 young in each brood. Substantial populations may be built up for they are able to withstand both a moderately high flowrate, provided they have a foothold, and low flow conditions, where they are more able than most to survive locally low dissolved oxygen concentrations which may occur in dead-legs.

The appearance of *Asellus* at consumers' taps is often the result of a change in flowrate, causing a dislodgement of the animal from the walls. Such an interruption may occur during the repair of a main, a mains flushing, intermittent service reservoir replenishment etc as well as the more localised flow changes associated with natural draw-off by consumers. There may be a significant population crash following a sudden reduction in supply of an otherwise continual resource, for example a prolonged algal bloom. Dead animals will be passively taken downstream as they have no predators in this environment. The partially decomposed animal fragments are thus able to pass through consumers' taps.

Gammarus and *Crangonyx*

These crustaceans are laterally compressed with curved backs, and are able to swim in moderate currents as well as lying in quiescent crevices in the walls of the mains. *Gammarus pulex* may reach 24 mm, although the more common *Crangonyx pseudogracilis* is only 10 mm in length. They breed in the mains, producing fewer young than *Asellus* but having more broods. Both feed on suspended particles and detritus, hence, as with *Asellus*, may reach high populations if large quantities of algae are allowed to pass through the treatment works.

Cyclops and other water fleas

The copepod water fleas of the *Cyclops* genus are very small, between 0.5 and 3 mm in length, depending on the species and sex of individuals. Of those that breed in the mains, the majority are carnivorous, feeding on a variety of organisms including smaller, herbivorous *Cyclops* species. They produce large numbers of young, and are noticed by a consumer because of their short, jerky movements.

Infestations may also be caused by the adults themselves passing through the treatment works in high numbers. This applies also for certain other copepods, for example the harpacticoids, and the cladocerans *Chydorus* and *Alona*, though other cladocera such as *Bosmina* and *Daphnia* are usually retained by treatment as are the planktonic, herbivorous copepods such as *Diaptomus*.

Nais

Naids are very thin, white worms, which may reach 25 mm in length, noticeable because of the continuous, coiling motion of their bodies. In warm temperatures, asexual reproduction occurs rapidly, with chains of individuals being formed. Naid worms are more susceptible than most other groups to chlorination.

Chironomid larvae

Many flying insects have aquatic pre-adult stages, and the non-biting midges (*Chironomidae*) are typical. The eggs are laid in gelatinous strings from which hatch tiny larvae. These pass through several moults before attaining full larval size, which may exceed 25 mm in length, depending on species, with a whitish, greenish or bright red coloration. Since the larval stage of chironomids may last two years or more, if large numbers of the early stages pass through the treatment works, these may concentrate in particular areas and build up numbers locally. Most species build a protective tube of organic and inorganic material, which adheres to the pipe wall.

Animal fragments in discoloured water

The majority of complaints of discoloured water are not due to animalcules but to vegetable material which persists throughout the treatment process or else to inorganic fragments such as manganese or iron. However, the most common complaint of 'black bits' caused by animalcules is due to *Asellus* frass, which if present in large quantities is highly indicative of an infestation. The pellets are of a very characteristic shape, thus easily identified. The cast exoskeletons of certain crustaceans, especially *Alona* and *Chydora* species (both 'water fleas') may become discoloured with iron which makes them very much more visible. High numbers of these may cause a 'dirty water' complaint. Occasionally, large numbers of dead diatoms (with brown, silicaceous shells) have been responsible for discoloured water—the cause often a stirring up of a deposit such as in a service reservoir. The algae, having passed through the treatment works, cannot survive in the absence of light, and being particularly dense, settle out readily in the reservoir.

Summary

Consumer complaints related to animalcules are due either to whole, visible animals, animal fragments or 'dirty' water. In general, lowland surface waters produce the greatest numbers of animalcules. Passage through a treatment works is the most likely point of entry into the mains but other key points are service reservoirs and during mains repair work. Once in the mains system, the animals colonise deposits and interior surfaces of pipes. Alternatively, they may be free-swimming in the water itself. Only a limited number of organisms are able to breed successfully in the distribution system; these include *Asellus*, the freshwater shrimps, naid worms and some *Cyclops* and *Alona* species. Chironomid midge larvae, whilst not breeding in the mains may nevertheless populate the system in large numbers through constant recruitment through the treatment works. Chironomids are ubiquitous in surface waters.

Heavy infestations of animals generally occur in 'dead ends' of mains, where the water flow within the mains is very low, or in service reservoirs where there is a slow water turnover. In these situations, the build-up of organic matter from droppings and dead animals can lead to an increase in chlorine demand, a deterioration in microbiological quality of the water and cause taste and odour complaints.

In attempting to identify the source of infestation/point of entry of the animalcule it may often be necessary to use microscopy to identify the species precisely. Otherwise, only generalisation is possible.

Measures to control infestations include the application of approved disinfestation agents, systematic flushing, and swabbing or air-scouring the affected areas.

As mentioned above, the presence of these animals in a water supply may indicate a failure at the treatment works or a breach in the integrity of the distribution system, and the detection of animals in a potable water must always lead to an immediate assessment of microbiological quality. However, there are no known health risks associated with these animalcules.

Appendix G: Other possible faecal indicator organisms

The *Bacteroides fragilis* group

This group consists of five separate species—*B. fragilis*, *B. vulgatus*, *B. thetaiotaomicron*, *B. ovatus* and *B. diastonis* (Cato and Johnson 1976). They are Gram-negative, obligately anaerobic, non-motile, non-sporing bacteria which have been found in the faeces of humans and domestic and farm animals (Allsop and Stickler 1984). They outnumber *E. coli* in the faeces of humans, cats and dogs but the converse has been found for farm animals and seagulls. While they have potential as indicators of recent human faecal pollution they do not appear to have any advantage over *E. coli* for, although their resistance to chlorine is similar, their survival in water is relatively poor (Allsop and Stickler 1984).

Isolation and enumeration are by membrane filtration using Wilkins and Chalgren's agar containing gentamicin and penicillin G with anaerobic incubation at 30°C for 4 hours followed by 44 hours at 37°C (Allsop and Stickler 1984).

***Bifidobacterium* species**

The genus *Bifidobacterium* comprises many species all of which are Gram-positive, obligately anaerobic, catalase-negative bacteria which attack sugars fermentatively. They are rod shaped but highly variable in appearance showing Y- or V-shaped bifurcated forms. Bifidobacteria have been found in the faeces of humans, pigs, cattle and sheep but not in horses, cats, dogs, poultry, goats or beavers (Mara and Oragui 1983, Resnick and Levin 1981). Sorbitol-fermenting strains (*B. adolescentis* and *B. breve*) have been found only in humans and could have potential as indicators of recent human faecal pollution of water. However, the use of this group as indicators of faecal pollution is limited by poor survival and sensitivity to chlorine relative to *E. coli* (Resnick and Levin 1981).

Isolation and enumeration of bifidobacteria are by membrane filtration using YN-17 medium and sorbitol-fermenting strains by using Human Bifid Sorbitol Agar (Mara and Oragui 1983). Media are incubated anaerobically at 37°C for 48 hours.

Candida albicans

Candida albicans is a typical yeast in culture. Cells are generally ovoid but not always of a characteristic shape. Asexual reproduction is by bi-polar budding, multiple buds may be formed at either or both poles; no unique pigment is produced and carbohydrate media are fermented.

C. albicans has been found in 14–31% of human faeces as compared to 87–100% for total coliforms including *E. coli* (Geldreich 1977). This relatively low incidence and the low numbers found in polluted waters when compared with faecal coliforms do not commend its use as an indicator of faecal pollution in treated water.

Rhodococcus coprophilus

R. coprophilus is an aerobic nocardia-like actinomycete which is excreted by farm animals but not humans. It can survive in water for long periods and could have potential for the detection of faecal pollution by farm animals, but isolation is slow (17–18 days) and difficult in unskilled hands. It is considered to be of value only in untreated waters as little is known about its resistance to chlorine (Rowbotham and Cross 1977, Mara and Oragui 1981). Isolation and enumeration are carried out using modified M3 agar (Mara and Oragui 1981).

Coliphages

Coliphages have been proposed as indicators of water pollution and as possible models for enteroviruses during treatment of drinking water (Stetler 1984, IAWPRC 1991). It has also been reported that coliphages and enteroviruses are removed or inactivated at similar rates during treatment processes, and that certain coliphages are at least as resistant to both chlorination and environmental stresses as enteroviruses (Kott et al 1974, Kott, Ben-Ari and Vinokur 1978, Simkova and Cervenka 1981, Havelaar and Hogeboom 1984). Stetler (1984) reported that enterovirus isolates correlated better with coliphage counts than with total coliforms, faecal coliforms, faecal streptococci or standard plate count organisms in analytical data from raw water. These are important considerations, as it is reported that human enteric viruses may survive water treatment processes, especially chlorination, better than pathogenic or indicator bacteria (Berg et al 1978). Both coliphages and enteroviruses have been found in chlorinated water (Stetler 1984).

Coliphages appear to be widely and consistently distributed in polluted surface waters, sewage and effluents. However, there is some uncertainty as to the ubiquity of coliphages as faecal rather than sewage indicators (Parry 1980) and it has been suggested that some types of coliphage are carried by only a small proportion of the human population (Havelaar, Furuse and Hogeboom 1986). If this is so, then coliphages would not be good indicators of faecal pollution but might still have real value in the assessment of treatment efficacy in removing or inactivating viruses, since they are routinely present in sewage-contaminated raw waters.

The term 'coliphages' describes a range of bacterial viruses which can be divided into two main groups. These are the 'somatic' phages which gain entry to the bacterial cell through its cell wall, and the 'F-specific' or 'male-specific' phages which attach to the sex pili of the *E. coli* host. Both somatic and F-specific RNA bacteriophages can multiply in an appropriate host in sewage, but F-specific RNA bacteriophages are thought less likely to multiply in unpolluted waters. F-specific coliphages have similar resistance to some important human viruses for environmental stresses, including disinfection. Consequently, it has been argued that the presence of F-specific RNA bacteriophages is a better index of sewage pollution and the possible presence of human viruses than somatic coliphages (IAWPRC 1991). However, because RNA coliphages are frequently absent from faeces, their absence from water does not necessarily indicate the absence of human viruses (IAWPRC 1991).

The most sensitive approach to the isolation of somatic phages appears to be to select a host strain of *E. coli* which is sensitive to the widest range of phage types. *E. coli* CN seems to be particularly suitable (Grabow and Coubrough 1986, Havelaar, Furuse and Hogeboom 1986).

Coliphage isolation techniques are straightforward, inexpensive and rapid, requiring overnight incubation only, and coliphage plaques may often be counted after as little as eight hours (Grabow and Coubrough 1986). Coliphage isolation may therefore prove to be a valuable monitoring tool for evaluating the effectiveness of treatment processes, especially with respect to removal or inactivation of enteroviruses. Lowland river sources are particularly likely candidates for this type of monitoring.

References

ACDP (1990) Categorisation of pathogens according to hazard and categories of containment, 2nd edition, 1990. Advisory Committee on Dangerous Pathogens. London, HMSO.

ALLSOP, K. & STICKLER, D.J. (1984) The enumeration of *Bacteroides fragilis* group organisms from sewage and natural waters. Journal of Applied Bacteriology, **56**, 15–24.

ANON (1978) A contaminated hospital water supply. Communicable Disease Report (48), 2. London, Internal Publication of the Public Health Laboratory Service.

ANON (1979) Contaminated water supply in a new hospital. Communicable Disease Report (1), 2. London, Internal Publication of the Public Health Laboratory Service.

APHA (1989) Standard methods for the examination of water and waste water, 15th edition. Washington D.C., American Public Health Association.

ASHWORTH, J. & COLBOURNE, J.S. (1987) The testing of non-metallic materials for use in contact with potable water, and the inter-relationships with in-service use. In: Industrial Microbiological Testing, (Editors, Hill A. & Hopton J.). Society for Applied Bacteriology Technical Series No. 23, 151–170. London, Academic Press.

AVAB (1987) Drinks Vending Machines: Code of Practice on Hygiene and Water Supply. Banstead, Automatic Vending Association of Britain.

AVAB (1989) Microbiological Recommendations for Drinks Vending Machines. Banstead, Automatic Vending Association of Britain.

BENTON, C., FORBES, G.I., PATERSON, G.M., SHARP, J.C.M. & WILSON, T.S. (1989) The incidence of waterborne and water-associated disease in Scotland from 1945 to 1987. Water Science and Technology, **21**, 125–129.

BERG, G., DAHLING, D.R., BROWN, G.A. & BERMAN, D. (1978) Validity of faecal coliforms and faecal streptococci as indicators of viruses in chlorinated primary sewage effluents. Applied and Environmental Microbiology, **36**, 880–884.

BOLTON, F.J., COATES, D., HINCHCLIFFE, P.M. & ROBERTSON, L. (1983) Comparison of selective media for isolation of *Campylobacter jejuni/coli*. Journal of Clinical Pathology, **36**, 78–83.

BOLTON, F.J., HUTCHINSON, D.M. & COATES, D. (1986) Comparison of three selective agars for isolation of *Campylobacter*. European Journal of Clinical Microbiology, **5**, 466–468.

BOLTON, F.J. & ROBERTSON, L. (1982) A selective medium for isolating *Campylobacter jejuni/coli*. Journal of Clinical Pathology, **35**, 462–467.

BONDE, G.J. (1977) Bacterial indicators of water pollution. Advances in Aquatic Microbiology, **1**, 273–364.

BR/IEHO (1985) The supply of water for train catering. The Report of a Working Party. London, British Rail Board and the Institution of Environmental Health Officers.

BROWN, M.W.R., & FOSTER, J.H.S. (1970) A simple diagnostic milk medium for *Pseudomonas aeruginosa*. Journal of Clinical Pathology, **23**, 172–177.

BSI (1961) British Standard 3421: Specification for performance of electrically-heated sterilizing ovens. London, British Standards Institution.

BSI (1976) British Standard 700: Graduated pipettes. London, British Standards Institution.

BSI (1983) British Standard 6297: Code of practice for design and installation of small sewage treatment works. London, British Standards Institution.

BSI (1988) British Standard 6920: Suitability of non-metallic products for use in contact with water intended for human consumption with regard to their effect on the quality of the water. London, British Standards Institution.

BSI (1992a) British Standard 7592: Methods for sampling for Legionella organisms in water and related materials. London, British Standards Institution.

BSI (1992b) Draft for Development DD 211. Method for detection and enumeration of Legionella organisms in water and related materials. London, British Standards Institution.

BSI (1993) British Standard 2646: Autoclaves for sterilization in laboratories, Part 1—Specification for design, construction and performance; Part 3—Guide to safe use and operation; Part 5—Methods of test for function and performance. London, British Standards Institution.

BURMAN, N.P. (1955) The standardisation and selection of bile salt and peptone for culture media used in the bacteriological examination of water. Proceedings of the Society for Water Treatment and Examination, **4**, 10–26.

BUTTIAUX, R. (1958) Les streptococques fecaux des intestins humains et animaux. Annales de l'Institut Pasteur, **94**, 778–782.

CATO, E.P. & JOHNSON, J.L. (1976) Re-instatement of species rank for *Bacteroides fragilis*, *Bacteroides ovatus*, *Bacteroides diastonis*, *Bacteroides thetaiotaomicron* and *Bacteroides vulgatis*: Designation of neotype strains for *Bacteroides fragilis* (Veillon and Zeiber) Castellani and Chalmers, and *Bacteroides thetaiotaomicron* (Diastas) Castellani and Chalmers. International Journal of Systematic Bacteriology, **26**, 230–237.

CHAPMAN, P.A., SIDDONS, C.A., WRIGHT, D.J., NORMAN, P., FOX, J., CRICK, E. (1993). Cattle as a possible source of verocytotoxin-producing *Escherichia coli* 0157:H7 infections in man. Epidemiology and Infection, **111**, 439–447.

CLARKE, P.H. (1953) Growth of streptococci in a glucose phenolphthalein broth. Journal of General Microbiology, **9**, 350–352.

COCHRAN, W.G. (1950) Estimation of bacterial densities by means of "the most probable number". Biometrics, **6**, 105–116.

COLBOURNE, J.S. (1985) Materials usage and their effects on the microbiological quality of water supplies. Journal of Applied Bacteriology Symposium Supplement, **59**, 47S-58S

COLLINS, C.H. (1989) Collins and Lyne's Microbiological Methods, 6th edition. (Editors, Collins C.H., Lyne P.M. & Grange J.M.). London, Butterworths.

COLLINS, C.H., GRANGE, J.M. & YATES, M.D. (1984) A review—mycobacteria in water. Journal of Applied Bacteriology, **57**, 193–211.

COLLINS, C.H., HARTLEY, E.G. & PILSWORTH, R. (1974) The prevention of laboratory acquired infection. London, HMSO (PHLS Monograph Series No. 6).

COWAN, S.T. (1993) Cowan and Steels' Manual for the Identification of Medical Bacteria, 3rd edition. (Editors, Barrow G.I. & Feltham R.K.A.). London, Cambridge University Press.

COX, C.R. (1964) Operation and control of water treatment processes. Geneva, World Health Organization (WHO Monograph Series No. 49).

CROSSLEY, E.L. (1941) The routine detection of certain sporforming anaerobic bacteria in canned foods. Journal of the Society of Chemical Industry, **60**, 131–136.

DH (1988) HC(88)64. Health of the Population: Responsibilities of Health Authorities. London, Department of Health.

DH (1993a) Drinking water supplies: a microbiological perspective, (Editors, Dawson A. & West P.). Department of Health. London, HMSO.

DH (1993b) Annual Report of the Committees on Toxicity, Mutagenicity and Carcinogenicity of Chemicals in Food, Consumer Products and the Environment. London, HMSO 1993.

DH/WO (1988) The Control of Legionellae in Health Care Premises -A Code of Practice. Department of Health and Welsh Office, London, HMSO.

DHSS/WO (1976) Technical Memorandum No. 27. Cold Water Supply Storage and Mains Distribution. London, Department of Health and Social Security and Welsh Office.

DoE (1989) Code of Practice for the Agricultural Use of Sewage Sludge. Department of the Environment, London, HMSO.

DoE/DH (1990) Cryptosporidium in water supplies (Chairman of Working Party: Sir John Badenoch). Department of the Environment and Department of Health. London, HMSO.

DoE/WO/SO (1993) Manual on Treatment of Private Water Supplies, London, HMSO.

DRAKE, C.H. (1966) Evaluation of culture media for the isolation and enumeration of *Pseudomonas aeruginosa*. Health Laboratory Science, **3**, 10–19.

DT (1986) Recommendations to prevent contamination of ships freshwater storage and distribution systems. Merchant Shipping Notice No. M. 1214. London, Department of Transport.

DWI 1992—*Cryptosporidium* in Water Supplies—Progress with the National Research Programme. DWI 1990

DWI 1994. Information Letter 1/94, "Sampling from service reservoirs"

EDEL, W. & KAMPELMACHER, E.H. (1973) Comparative studies on the isolation of 'sub-lethally injured' *Salmonellae* in nine European laboratories. Bulletin of the World Health Organization, **48**, 167–174

EDWARDS, P.R. & FIFE, M.A. (1961) Lysine-iron agar in the detection of Arizona cultures. Applied Microbiology, **9**, 478–480

EISENHART, C. & WILSON, P.W. (1943) Statistical methods and control in bacteriology. Bacteriological Reviews, **7,** 57–137

FENG, P.C.S. & HARTMAN, P.A. (1982) Fluorogenic assay for immediate confirmation of *Escherichia coli*. Applied and Environmental Microbiology, **43**, 1320–1329

GELDREICH, E.E., (1977) Bacterial population and indicator concepts in faeces, sewage, stormwater and solid wastes. In: Indicators of Virus in Water and Food, 51–97. (Editor Berg, G). Ann Arbor, Michigan, Ann Arbor Sciences.

GELDREICH, E.E., TAYLOR, R.H., BLANNON, J.C. & REASONER, D.J. (1985) Bacterial colonization of point-of-use water treatment devices. Journal of the American Waterworks Association, **77**, 72–80.

GEORGE, H.A., HOFFMAN, P.S., SHIBERT, R.M. & KREIG, N.R. (1978) Improved media for growth and aerotolerance of *Campylobacter fetus*. Journal of Clinical Microbiology, **8**, 36–41.

GIBBS, B.M. & FREAME, B. (1965) Methods for the recovery of Clostridia from foods. Journal of Applied Bacteriology, **28**, 95–111.

GRABOW, W.O.K. & COUBROUGH, P. (1986) Practical direct plaque assay for coliphages in 100 mL samples of drinking water. Applied and Environmental Microbiology, **52**, 430–433.

GRAY, R.D. (1964) An improved formate lactose glutamate medium for the detection of *Escherichia coli* and other coliform organisms from water. Journal of Hygiene, **62**, 495–508.

GREENWOOD, M.H. (1993) Comparison of enrichment at 9°C and 21°C for recovery of *Yersinia* species from food and milk. Food Microbiology, **10**, 23–30.

GUTHOF, D. (1957) Streptokoken und Dysbakterie-Problem. Zentralblatt fur Bakteriologie, Parasitenkunde, Infektionskrankheiten und Hygiene, Erste Abteilung, Originale, **170**, 327–333.

HALL, T. & HYDE, R. (1992) Water Treatment Processes and Practice. Swindon, Water Research Centre.

HAMMARSTROM, E. & LJUTOV, V. (1954) Concentration technique for demonstrating small amounts of bacteria in tap-water. Acta Pathologica et Microbiologica Scandinavica, **35**, 365–369.

HANDFORD, P.M. (1974) A new medium for the detection and enumeration of *Clostridium perfringens* in foods. Journal of Applied Bacteriology, **37**, 559–570.

HARMON, P.L., STUART, P. & MAWER, S.L. (1991) Cefsulodin as a selective agent for the isolation of Salmonella from materials heavily contaminated with *Pseudomonas* species. PHLS Microbiology Digest, **8**, 67–69.

HARMON, S.M., KAUTTER, D.A. & PEELER, J.T. (1971) Improved medium for the enumeration of *Clostridium perfringens*. Applied Microbiology, **27**, 688–692.

HAUSCHILD, A.H.W. & HILLSHEIMER, R. (1974) Enumeration of foodborne *Clostridium perfringens* in egg yolk free tryptose-sulphite-cycloserine agar. Applied Microbiology, **27**, 521–526.

HAVELAAR, A.H., DURING, M. & VERSTEEGHT, J.F.M. (1987) Ampicillin-dextrin agar medium for the enumeration of *Aeromonas* species in water by membrane filtration. Journal of Applied Bacteriology, **62**, 279–287.

HAVELAAR, A.H., FURUSE, K. & HOGEBOOM, W.M. (1986) Bacteriophages and indicator bacteria in human and animal faeces. Journal of Applied Bacteriology, **60**, 255–262.

HAVELAAR, A.H. & HOGEBOOM, W.M. (1984) A method for the enumeration of male-specific bacteriophages in sewage. Journal of Applied Bacteriology, **56**, 439–447.

HOBBIE, J.E., DALEY, R.J. & JASPER, S. (1977) Use of Nucleopore filters for counting bacteria by fluorescence microscopy. Applied and Environmental Microbiology, **33**, 1225–1228.

HOBBS, B.C. & ALLISON, V.D. (1945a) Studies on the isolation of *Bact. typhosum* and *Bact. paratyphosum B*. Part I. Monthly Bulletin of the Ministry of Health and Public Health Laboratory Service, **4**, 12–19.

HOBBS, B.C. & ALLISON, V.D. (1945b) Studies on the isolation of *Bact. typhosum* and *Bact. paratyphosum B*. Part III. Monthly Bulletin of the Ministry of Health and the Public Health Laboratory Service, **4**, 63–68.

HOLDEN, W.S. ed. (1970) Water treatment and examination. London, Churchill.

HSAC (1991a) Safety in Health Service Laboratories—Safe working and the prevention of infection in clinical laboratories. Health Services Advisory Committee. London, HMSO.

HSAC (1991b) Safety in Health Service Laboratories—Safe working and the prevention of infection in clinical laboratories—model rules for staff and visitors. Health Services Advisory Committee. London, HMSO.

HSE (1991) The control of legionellosis including Legionnaires' disease. Health and Safety Executive Series Booklet HS(G)70. London, HMSO.

HUGH, R. & LEIFSON, E. (1953) The taxonomic significance of fermentative versus oxidative metabolism of carbohydrates by various Gram negative bacteria. Journal of Bacteriology, **66**, 24–26.

HURLEY, M.A. & ROSCOE, M.E. (1983) Automated statistical analysis of microbial enumeration by dilution series. Journal of Applied Bacteriology, **55**, 159–164.

HUTCHINSON, M. & RIDGWAY, J.W. (1977). Microbiological Aspects of Drinking Water Supplies. In: Aquatic Microbiology (Editors Skinner F.A. & Shewan J.M.). Society for Applied Bacteriology Symposium Series No.6, 179–218. London, Academic Press.

IAMS (1958) Report of the Enterobacteriaceae Subcommittee of the Nomenclature Committee of the International Association of Microbiological Societies. International Bulletin of Bacterial Nomenclature and Taxonomy, **8**, 25–70.

IARC (1991) Monograph on the Evaluation of Carcinogenic Risks to Humans, Volume 532—Chlorinated Drinking Water; chlorination by-products; some other halogenated compounds; cobalt and cobalt compounds. Lyon, World Health Organization International Agency for Research on Cancer.

IAWPRC (1991) Bacteriophages as model viruses in water quality control. International Association for Water Pollution Research and Control Study Group on Health Related Water Microbiology. Water Research, **25**, 529–545.

ICMSF (1978) Micro-organisms in foods 1. Their significance and methods of enumeration, 2nd edition. International Commission on Microbiological Specification for Foods. Toronto, University of Toronto Press.

ISO (1986) Quality-Vocabulary (ISO/IEC Standard 8402). International Organization for Standardization.

KAMPFER, P., RAUHOFF, O. & DOTT, W. (1991) Glycosidase profiles of members of the family Enterobacteriaceae. Journal of Clinical Microbiology, **29**, 2877–2879.

KAPER, J.B., SEIDLER, R.J., LOCKMAN, H. & COLWELL, R.R. (1979) Medium for the presumptive identification of *Aeromonas hydrophila* and Enterobacteriaceae. Applied and Environmental Microbiology, **38**, 1023–1026.

KING, S., & METZGER, W.I. (1968) A new plating medium for the isolation of enteric pathogens I. Hektoen enteric agar. II. Comparison of Hektoen enteric agar with SS and EMB agar. Applied Microbiology, **16**, 577–581.

KOBAYASHI, T., ENOMOTO, S., SAKAZAKI, R. & KUWAHARA, S. (1963) A new selective medium for pathogenic vibrios, TCBS agar (modified Nakanishi's agar). Japanese Journal of Bacteriology, **18**, 387–411.

KOTT, Y., BEN-ARI, H., & VINOKUR, L. (1978) Coliphages survival as viral indicator in various wastewater quality effluents. Progress in Water Technology, **10**, 337–346.

KOTT, Y., ROZE, N., SPERBER, S. & BETZER, N. (1974) Bacteriophages as viral pollution indicators. Water Research, **8**, 165–171.

KOVACS, N. (1928) Eine vereinfachte Methods zum Nachweis der Indolbildung durch Bakterien. Zeitschrift fur Immunitatsforschung und experimentelle Therapie, **55**, 311–315.

LANDER, K. & GILL, K.P.W. (1985) Campylobacters. In: Isolation and Identification of Micro-organisms of Medical and Veterinary Importance (Editors, Collins C.H. & Grange J.M.). Society for Applied Bacteriology Technical Series No. 6, 123–142. London, Academic Press.

LECHEVALLIER, M.W., CAWTHON, C.D. & LEE, R.G. (1988) Inactivation of biofilm bacteria. Applied and Environmental Microbiology, **54**, 2492–2499.

Le MINOR, L., ROHDE, R. & TAYLOR, J. (1970) Nomenclature des Salmonella. Annales Institut Pasteur, Microbiologie **119**, 206–210.

LIGHTFOOT, N.F., TILLETT, H.E., BOYD, P. & EATON, S., (1994) Duplicate split samples for internal quality control in routine water microbiology. Letters in Applied Microbiology, **19**.

MAFF/WOAD (1985) Code of Good Agricultural Practice. London, Ministry of Agriculture, Fisheries and Food and Welsh Office Agriculture Department.

MAIR, N.S., & FOX, E. (1986) Yersiniosis: Lab diagnosis, Clinical Features, Epidemiology. London, Public Health Laboratory Service.

MALLMAN, W.L. & SELIGMANN, E.B. (1950) A comparative study of media for the detection of streptococci in water and sewage. American Journal of Public Health, **40**, 286–289.

MAN, J.C. de (1975) The probability of most probable numbers. European Journal of Applied Microbiology, **1**, 67–78.

MARA, D.D. & ORAGUI, J.I. (1981) Occurrence of *Rhodococcus coprophilus* and associated actinomycetes in faeces, sewage and freshwater. Applied and Environmental Microbiology, **42**, 1037–1042.

MARA, D.D. & ORAGUI, J.I. (1983) Sorbitol fermenting bifidobacteria as specific indicators of human faecal pollution. Journal of Applied Bacteriology, **55**, 349–358.

MARA, D.D. & ORAGUI, J.I. (1985) Bacteriological methods for distinguishing between human and animal faecal pollution of water: results of field work in Nigeria and Zimbabwe. Bulletin of the World Health Organization, **63**, 773–783.

McCOY, J.H. & SPAIN, G.E. (1969) Bismuth sulphite media in the isolation of Salmonellae. In: Isolation Methods for Microbiologists (Editors, Shapton D.A. & Gould G.W.). Society for Applied Bacteriology Technical Series No. 3, 17–27. London, Academic Press.

McCRADY, M.H. (1915) The numerical interpretation of fermentation-tube results. Journal of Infectious Diseases, **17**, 183–212.

McCRADY, M.H. (1918) Tables for rapid interpretation of fermentation-tube results. Public Health Journal, Toronto, **9**, 201–220.

MEAD, G.C. (1964) Isolation and significance of *Streptococcus faecalis* sensu strictu. Nature, London, **204**, 1224–1225.

MEAD, G. C., ADAMS, B. W., (1977) A selective medium for the rapid isolation of Pseudomonads associated with poultry meat spoilage. British Poultry Science, **18**, 661–670.

MOORE, B. (1948) The detection of paratyphoid carriers in towns by means of sewage examination. Monthly Bulletin of the Ministry of Health and Public Health Laboratory Service, **7**, 241–248.

MOORE, B. (1950) The detection of typhoid carriers in towns by means of sewage examination. Monthly Bulletin of the Ministry of Health and Public Health Laboratory Service, **9**, 72–78.

MOORE, B., PERRY, E.L. & CHARD, S.T. (1952) A survey by the sewage swab method of latent enteric infection in an urban area. Journal of Hygiene, **50**, 137–156.

MOSSEL, D.A.A. (1982) Microbiology of foods: The ecological essentials of assurance and assessment of safety and quality. Utrecht, University of Utrecht.

NRA (1992) Policy and Practice for the Protection of Groundwater. Bristol, National Rivers Authority.

NWC (1983) Safety in microbiology laboratories in the water industry, by the NWC Microbiological Working Group. London, National Water Council.

ORAGUI, J.I. & MARA, D.D. (1981) A selective medium for the enumeration of *Streptococcus bovis* by membrane filtration, Journal of Applied Bacteriology, **51**, 85–93.

PARRY, O.T. (1980) Bacteriophages as water pollution indicators. Ph.D. Thesis, Liverpool Polytechnic Department of Biology.

PAYMENT, P., FRANCO, E., RICHARDSON, L. & SIEMIATYCHI, J. (1991) Gastrointestinal health effects associated with the consumption of drinking water produced by point-of-use domestic reverse osmosis filtration units. Applied and Environmental Microbiology, **57**, 945–948.

PAYMENT, P. (1991a) Elimination of coliphages, *Clostridium perfringens* and human enteric viruses during drinking water treatment: results of large-volume samplings. Water Science and Technology, **24**, 213–215.

PAYMENT, P. (1991b) Fate of human enteric viruses, coliphages and *Clostridium perfringens* during drinking water treatment. Canadian Journal of Microbiology, **37**, 154–157.

PECD 7/7/413 A report published by WRc entitled "Effects of storage on analysis results for the total and faecal coliform parameters", No 3705.

PECD 7/7/424 "The evaluation of presence-absence test-kits for coliform organisms".

PHLS (1953) The effect of sodium thiosulphate on the coliform and *Bacterium coli* counts of non-chlorinated water samples, by the Public Health Laboratory Service Water Subcommittee. Journal of Hygiene, **51**, 572–577.

PHLS (1969) A minerals-modified glutamate medium for the enumeration of coliform organisms in water, by the Public Health Laboratory Service Standing Committee on the Bacteriological Examination of Water Supplies. Journal of Hygiene, **67**, 367–374.

PHLS (1978) Waterborne infectious disease in Britain, by the Public Health Laboratory Service Standing Sub-Committee on the Bacteriological Examination of Water Supplies. Journal of Hygiene, **81**, 139–149.

PHLS (1992) Immunofluorescence: antigen detection techniques in diagnostic microbiology (Editor, Caul E.O.). London, Public Health Laboratory Service.

PHLS (1993) Safety procedures: notes for guidance, 4th edition. London, Public Health Laboratory Service.

PHLS/SCA (1980a) Membrane filtration media for the enumeration of coliform organisms and *Escherichia coli* in water: comparison of Tergitol 7 and lauryl sulphate with Teepol 610, by a Joint Committee of the Public Health Laboratory Service and Standing Committee of Analysts. Journal of Hygiene, **85**, 181–191.

PHLS/SCA (1980b) A comparison between minerals-modified glutamate medium and lauryl tryptose lactose broth for the enumeration of *Escherichia coli* and coliform organisms in water by the multiple tube method, by a Joint Committee of the Public Health Laboratory Service and the Standing Committee of Analysts. Journal of Hygiene, **85**, 35–49.

PHLS/SCA (1980c) Single tube confirmatory tests for *Escherichia coli*, by a Joint Committee of the Public Health Laboratory Service and the Standing Committee of Analysts, Journal of Hygiene, **85**, 51–57.

REASONER, D.J., & GELDREICH, E.E. (1985) A new medium for the enumeration and subculture of bacteria from potable water. Applied and Environmental Microbiology, **49**, 1–7.

RESNICK, I.G., & LEVIN, M.A. (1981) Assessment of bifidobacteria as indicators of human faecal pollution. Applied and Environmental Microbiology, **42**, 433–438.

RICHARDSON, A.J., FRANKENBERG, R.A., BUCK, A.C., SELKON, J.B., COLBOURNE, J.S., PARSONS, J.S. & MAYON-WHITE, R.T. (1991) An outbreak of waterborne cryptosporidiosis in Swindon and Oxfordshire. Epidemiology and Infection, **107**, 485–495.

RIPPEY, S.R. & CABELLI, V.J. (1979) Membrane filter procedure for enumeration of *Aeromonas hydrophila* in fresh waters. Applied and Environmental Microbiology, **38**, 108–113.

ROWBOTHAM, T.J. & CROSS, T. (1977) Ecology of *Rhodococcus coprophilus* and associated actinomycetes in freshwater and agricultural habitats. Journal of General Microbiology, **100**, 231–240.

SARTORY, D.P. (1986) Membrane filtration enumeration of faecal clostridia and *Clostridium perfringens* in water. Water Research, **20**, 1255–1260.

SCA (1980) Chemical disinfecting agents in water and effluents, and chlorine demand 1980, by the Standing Committee of Analysts of the Department of the Environment and the National Water Council, London, HMSO. (Methods for the Examination of Water and Associated Materials).

SCA (1983) Methods for the isolation and identification of *Salmonellae* (other than *Salmonella typhi*) from water and associated materials 1982, by the Standing Committee of Analysts of the Department of the Environment and the National Water Council, London, HMSO. (Methods for the Examination of Water and Associated Materials).

SCA (1989) Isolation and identification of *Giardia* cysts and *Cryptosporidium* oocysts and free living pathogenic amoebae in water 1989, by the Standing Committee of Analysts of the Department of the Environment, London, HMSO. (Methods for the Examination of Water and Associated Materials).

SCA (1994) Methods for the isolation of human enteric viruses from water and related materials 1994, by the Standing Committee of Analysts of the Department of the Environment, London, HMSO. (Methods for the Examination of Water and Associated Materials).

SCHIEMANN D.A. (1982) Development of a two-step enrichment procedure for recovery of *Yersinia enterocolitica* from food. Applied and Environmental Microbiology, **43**, 14–27.

SHAHIDI, A. & FERGUSON, A.R. (1971) New quantitative, qualitative and confirmatory media for rapid analysing of food for *Clostridium perfringens*. Applied Microbiology, **21**, 500–506.

SHORT, C.S. (1988) The Bramham incident, 1980—an outbreak of waterborne infection. Journal of the Institution of Water and Environmental Management, **2**, 383–390.

SIMKOVA, A. & CERVENKA, J. (1981) Coliphages as ecological indicators of enteroviruses in various water systems. Bulletin of the World Health Organization, **59**, 611–618.

SKEAT, W.O. ed. (1969) Manual of British water engineering practice, 4th edition. Vol. III. London, Heffer.

SKIRROW, M.B. & BENJAMIN, J. (1980) Differentiation of enteropathogenic *Campylobacter*. Journal of Clinical Pathology, **33**, 1122.

SLANETZ, L.W. & BARTLEY, C.H. (1957) Numbers of enterococci in water, sewage and faeces determined by the membrane filter technique with an improved medium. Journal of Bacteriology, **74**, 591–595.

STENGREN, S.R. & STARZYK, M.J. (1984) A modified medium for the recovery of *Staphylococcus* from water. Microbios, **41**, 191–203.

STETLER, R.E. (1984) Coliphages as indicators of enteroviruses. Applied and Environmental Microbiology, **48**, 668–670.

SWAROOP, S. (1938) Numerical estimation of *B. coli* by dilution method. Indian Journal of Medical Research, **26**, 353–378.

SWAROOP, S. (1951) The range of variation of the most probable number of organisms estimated by the dilution method. Indian Journal of Medical Research, **39**, 107–134.

TABAQCHALI, S., STEVENS, J.K. & GAZIDES, C. (1977) Worms in hospital cold water. Lancet, **2**, 84–85.

TAYLOR, W.I. & HARRIS, B. (1965) Isolation of Shigellae I. Xylose lysine agars: new media for isolation of enteric pathogens. American Journal of Clinical Pathology, **44**, 471–475.

THOMAS, M.E.M., PIPER, E., & MAURER, I.M. (1972) Contamination of an operating theatre by Gram-negative bacteria. Examination of water supplies, cleaning methods and wound infections. Journal of Hygiene, **70**, 63–73.

THORNLEY, M.J. (1960) The differentiation of *Pseudomonas* from other Gram negative bacteria on the basis of arginine metabolism. Journal of Applied Bacteriology, **73**, 37–52.

TILLETT, H.E. (1986) Reporting bacteriological counts from water samples: how good is the information from an individual sample? In: Statistical Aspects of Water Quality Monitoring (Editors, El-Shaarawi A.H. & Kwiatkowski R.E.). Developments in Water Science, Volume 27, 221–230. Amsterdam, Elsevier Science Publishers.

TILLETT, H.E. (1987) Most probable numbers of organisms: revised tables for the multiple tube method. Epidemiology and Infection, **99**, 471–476.

TILLETT, H.E. (1993) Potential inaccuracy of microbiological counts from routine water samples. Water Science and Technology, **27**, 15–18.

TILLETT, H.E. & BENTON, C. (1993) Effects of transit time on indicator organism counts from water samples. PHLS Microbiology Digest, **10**, 116–117. London, Public Health Laboratory Service.

TILLETT, H.E. & COLEMAN, R. (1985) Estimated numbers of bacteria in samples from non-homogeneous bodies of water: how should MPN and membrane filtration results be reported? Journal of Applied Bacteriology, **59**, 381–388.

TILLETT, H.E. & FARRINGTON, C.P. (1991) Inaccuracy of counts of organisms in water or other samples: effects of pre-dilution. Letters in Applied Microbiology, **13**, 168–170.

VASSILIADIS, P. (1983) The Rappaport-Vassiliadis (RV) enrichment medium for the isolation of salmonellas: an overview. Journal of Applied Bacteriology, **54**, 69–76.

VASSILIADIS, P., PATERAKI, E., APAICONOMU, N., PAPADAKIS, J.A. & TRICHOPOULOS, D. (1976) Nouveau procede d'enrichissement de *Salmonella*. Annales de Microbiologie (Insitut Pasteur), **127B**, 195–200.

WAA (1984) Actions to minimise the effect of pollution incidents affecting river intakes for public water supplies. The Water Authorities' Association and the Water Companies' Association. London, Water Authorities' Association.

WAA (1985) Guide to the microbiological implications of emergencies in the water services. London, Water Authorities' Association.

WAA (1988) Operational guidelines for the protection of drinking water supplies: safeguards in the operation and management of public water supplies in England and Wales. London, Water Authorities' Association.

WAITE, W.M, (1991) Drinking water standards: a personal perspective. In: Proceedings of the UK Symposium on Health-related Water Microbiology (Editors, Morris, R., Alexander, L. M., Wyn-Jones, P., and Sellwood, J.) pp52–65. London, International Association for Water Pollution Research and Control.

WALSBY, A.E. (1975) Gas vacuoles. Annual Review of Plant Physiology, **26**, 427–439.

WHITE, G.C. (1985) The Handbook of Chlorination, 2nd edition. New York, Van Nostrand Reinhold.

WHO (1977) Guide to hygiene and sanitation in aviation (Editor, Baily J.). Geneva, World Health Organization.

WHO (1984) Guidelines for Drinking Water Quality. Vol. 2, Health Criteria and Other Supporting Information. Geneva, World Health Organization.

WHO (1993) Guidelines for Drinking Water Quality, Volume 1 Recommendations, Second edition. Geneva, World Health Organization.

WILSON, W.J. & BLAIR, E.M.McV. (1927) Use of glucose bismuth sulphite iron medium for the isolation of *B. typhosus* and *B. proteus*. Journal of Hygiene, **26**, 374–391.

WOODWARD, R.L. (1957) How probable is the most probable number? Journal of the American Waterworks Association, **49**, 1060–1068.

WRIGHT D.J., CHAPMAN P.A. & SIDDONS, C.A., (1994). Immunomagnetic separation as a method for isolating *Escherichia coli* 0157:H7 from food samples. Epidemiology and Infection, **113**, 31–39.

ZADIK, P.M., CHAPMAN, P.A., & SIDDONS, C.A., (1993). Use of tellurite for the selection of verocytotoxin *Escherichia coli* 0157:H7. Journal of Medical Microbiology, **39**, 155–158.

Index

A

B

C

D

E

F

G

H

I

J

K

L

M

N

O

P

Q

U

V

W

X

Y

Address for correspondence

Correspondence about this booklet should be addressed to:

The Secretary
Standing Committee of Analysts
Department of the Environment
Drinking Water Inspectorate
Romney House
43 Marsham Street
London
SW1P 3PY

Printed in the United Kingdom for HMSO
Dd298255 11/94 C6 G3396 10170